The Project Manager's Emergency Kit

The Project Manager's Emergency Kit

Ralph L. Kliem

S^t_L

ST. LUCIE PRESS

A CRC Press Company
Boca Raton   London   New York   Washington, D.C.

Library of Congress Cataloging-in-Publication Data

Kliem, Ralph L.
  The project manager's emergency kit / by Ralph L. Kliem.
      p. cm.
  ISBN 1-57444-333-X
  1. Project management. I. Title.

T56.8 .K54 2002
658.4′04--dc21   2002069717

This book contains information obtained from authentic and highly regarded sources. Reprinted material is quoted with permission, and sources are indicated. A wide variety of references are listed. Reasonable efforts have been made to publish reliable data and information, but the author and the publisher cannot assume responsibility for the validity of all materials or for the consequences of their use.

Neither this book nor any part may be reproduced or transmitted in any form or by any means, electronic or mechanical, including photocopying, microfilming, and recording, or by any information storage or retrieval system, without prior permission in writing from the publisher.

The consent of CRC Press LLC does not extend to copying for general distribution, for promotion, for creating new works, or for resale. Specific permission must be obtained in writing from CRC Press LLC for such copying.

Direct all inquiries to CRC Press LLC, 2000 N.W. Corporate Blvd., Boca Raton, Florida 33431.

Trademark Notice: Product or corporate names may be trademarks or registered trademarks, and are used only for identification and explanation, without intent to infringe.

Visit the CRC Press Web site at www.crcpress.com

© 2003 by CRC Press LLC
St. Lucie Press is an imprint of CRC Press LLC

No claim to original U.S. Government works
International Standard Book Number 1-57444-333-X
Library of Congress Card Number 2002069717
Printed in the United States of America 1 2 3 4 5 6 7 8 9 0
Printed on acid-free paper

Dedication

To Jan, Jenny, and Jessica Replogle

Preface

Being a project manager is one of the most challenging and rewarding experiences of one's professional life. It is challenging because there is always a shortage of just about everything, from time to people. In the midst of these shortages is the increasing pressure to deliver a product or service at the right moment to the right people in the right way while meeting the right standards. It is rewarding because of the opportunity to overcome those challenges by delivering a product or service in a manner that satisfies everyone who has a stake in the outcome of a project — the project manager, the team, the customer, senior management, and others.

Unfortunately, most projects do not overcome their challenges, and, if they do, everyone has a feeling of "Thank God, it's over." Of course, it does not have to be that way, which is why I wrote this book. As a project manager myself, I know that successful outcomes require good knowledge and reliable application of the tools, techniques, and principles of project management. This book provides everything you need to get a project off to a solid start, put it in cruise control, get it to its destination, and overcome any emergencies that arise along the way. It is truly your emergency kit to keep handy while maneuvering down that long road called the project life cycle. The contents of this book have worked successfully for me and other project managers.

How so? Here is what you get in this book. Over 200 tools, techniques, and principles are presented in alphabetical order. Each entry presents an overview, the goals to achieve, a list of the benefits and possible obstacles you will encounter, and, finally, steps for its application. You can reference the List of Figures following this Preface to find the proper tool, technique, or principle to apply in a particular situation. The matrix lists the four major functions of project management (planning, organizing, controlling, and leading) plus a miscellaneous column. Each function, in turn, is divided into four categories: cost, schedule, quality, and people. Running down the left column are the topics covered in the text of the book. Check marks in the individual cells indicate relevant topics for the function and corresponding category in which you are interested. For example, look up bar (Gantt) charts. Notice that this entry has check marks under the category "schedule" for both the planning and controlling functions. The matrix tells you when to use a particular tool — in this case, when planning or controlling a project. You can then refer to the entry in the book (for example, bar [Gantt] charts) to learn more about the topic. Or, better yet, if you are having a particular problem (for example, with bar charts during planning), you can refer to the relevant topic for ideas on how to overcome it.

Like all emergency kits, it is important to keep this one available throughout the life cycle of your project. That way when you need something, either to get off to a good start or to overcome an obstacle, you can simply refer to the applicable tool, technique, or principle and apply it in a way that helps you to arrive at your destination. Used properly, it can be the emergency kit that helps you deliver a project or service at the right moment to the right people in the right way while meeting the right standards. What is more, you will satisfy everyone who has a stake in the outcome of your project — yourself, the team, the customer, and senior management.

Happy travels.

Ralph Kliem, PMP
Practical Creative Solutions, Inc.

List of Figures

Figure 1	Affinity diagram.	2
Figure 2	Bar chart.	5
Figure 3	Breakeven analysis.	8
Figure 4	Cause-and-effect graph.	13
Figure 5	Chunking.	18
Figure 6	Communication diagram.	20
Figure 7	Core team.	28
Figure 8	Critical issues and action items log.	31
Figure 9	Decision table. (From Project Management Seminar presented by Practical Creative Solutions, Inc., 1996.)	34
Figure 10	Decision tree.	35
Figure 11	Dependency relationships. (From Project Management Seminar presented by Practical Creative Solutions, Inc., 1996.)	36
Figure 12	Early and late start and finish dates. (From Project Management Seminar presented by Practical Creative Solutions, Inc., 1996.)	39
Figure 13	Entity-relationship diagrams.	43
Figure 14	Fast tracking.	48
Figure 15	Fishbone diagram.	49
Figure 16	Functional hierarchy diagram.	52
Figure 17	Golden vs. iron triangle of project management.	57
Figure 18	Issue–action diagram.	65
Figure 19	Key contact listing.	67
Figure 20	Lag. (From Project Management Seminar presented by Practical Creative Solutions, Inc., 1996.)	69
Figure 21	Unleveled histogram.	82
Figure 22	Leveled histogram.	82
Figure 23	Managerial grid.	85
Figure 24	Hierarchy of needs.	87
Figure 25	Matrix structure. (From Project Management Seminar presented by Practical Creative Solutions, Inc., 1996.)	88
Figure 26	Task force structure. (From Project Management Seminar presented by Practical Creative Solutions, Inc., 1996.)	88
Figure 27	Mind mapping.	94
Figure 28	Arrow diagram. (From Project Management Seminar presented by Practical Creative Solutions, Inc., 1996.)	101
Figure 29	Precedence diagram. (From Project Management Seminar presented by Practical Creative Solutions, Inc., 1996.)	101
Figure 30	Neural net.	102
Figure 31	Objectives and their relationship to goals.	106

Figure 32	Organization chart.	107
Figure 33	Pareto analysis chart.	113
Figure 34	PDCA cycle.	114
Figure 35	PERT estimating technique. (From Project Management Seminar presented by Practical Creative Solutions, Inc., 1996.)	116
Figure 36	Project life cycles.	129
Figure 37	Approach for selecting software.	131
Figure 38	Project wall layout.	136
Figure 39	Responsibility matrix. (From Project Management Seminar presented by Practical Creative Solutions, Inc., 1996.)	145
Figure 40	Scattergram.	152
Figure 41	Scope creep.	154
Figure 42	Skills matrix.	158
Figure 43	Span of control. (From Project Management Seminar presented by Practical Creative Solutions, Inc., 1996.)	160
Figure 44	Statement of understanding. (From Project Management Seminar presented by Practical Creative Solutions, Inc., 1996.)	165
Figure 45	Statistical process control.	166
Figure 46	Top-down and bottom-up thinking. (From Project Management Seminar presented by Practical Creative Solutions, Inc., 1996.)	180
Figure 47	Typical work breakdown structure. (From Project Management Seminar presented by Practical Creative Solutions, Inc., 1996.)	194
Figure 48	Work breakdown structure by deliverables. (From Project Management Seminar presented by Practical Creative Solutions, Inc., 1996.)	195
Figure 49	Work breakdown structure by phase. (From Project Management Seminar presented by Practical Creative Solutions, Inc., 1996.)	196
Figure 50	Work breakdown structure by responsibility.	197
Figure 51	Workflow symbols.	198
Figure 52	Example of work flow.	199

Contents

A 1

Activity-Based Costing and Traditional Accounting 1
Affinity Diagram 2
Alternative Working Schedules 3

B 5

Bar (Gantt) Charts 5
Benchmarking 6
Body Language 7
Brainstorming 7
Breakeven Analysis 8
Budgeting 9

C 11

Capability Maturity Model 11
Capacity Planning 12
Categories of Change 12
Cause-and-Effect Graph 13
Change Board 14
Change Control 15
Change Implementation 16
Checkpoint Review Meeting 16
Chunking 17
Client 18
Collecting Statistics 19
Communication Diagram 20
Communications Process 21
Configuration Management 21
Conflict Resolution 22
Consultants and Contractors 23
Contingency Planning 24
Continuous Improvement 24
Contracts 25
Controlling 26
Controls 27
Core Team 28
Cost Analysis 29
Creativity 29

Critical Chain 30
Critical Issues and Action Items Log 31
Critical Path 32

D 33

Decision Making 33
Decision Tables 34
Decision Trees 35
Dependency Relationships 36

E 39

Early and Late Start and Finish Dates 39
Earned Value 40
E-Mail 41
Emotional Intelligence 42
Enneagram 42
Entity-Relationship Diagrams 43
Estimating 44

F 47

Facilitation 47
Fast Tracking 48
Fishbone Diagram 49
Forms 50
Forward and Backward Passes 51
Frameworks and Methodologies 51
Functional Hierarchy Diagram 52

G 55

Globalization of Projects 55
Goal 55
Golden vs. Iron triangle of Project Management 56
Groupthink 57

H 59

Herzberg Theory of Motivation 59
Heuristics 59

I 61

Imagineering 61
Information Life Cycle 61
Internal Rate of Return 62
Interviewing 63

Intuition 64
ISO 9000 64
Issue–Action Diagram 65

K 67

Key Contact Listing 67

L 69

Lag 69
Lateral and Vertical Thinking 70
Leadership Skills 71
Leadership: Communicating Skills 71
Leadership: Interpersonal Skills 72
Leadership: Modeling Skills 73
Leadership: Team Bonding Skills 73
Leading 74
Leading: Being Supportive 74
Leading: Communicating 75
Leading: Maintaining Direction 76
Leading: Making Effective Decisions 76
Leading: Motivating 77
Leading: Providing Vision 77
Leading: Using Delegation Properly 78
Learning Curve 78
Learning Style 79
Left and Right Brain Thinking 80
Lessons Learned 81
Leveling 81
Listening and Hearing 83
Logical and Physical Designs 84

M 85

Managerial Grid 85
Maslow Hierarchy of Needs 86
Matrix and Task Force Structures 87
Matrix 89
Mean, Median, and Mode 89
Meetings 90
Memo 91
Memorization 92
Mentoring vs. Coaching 92
Metrics 93
Mind Mapping 94
Modeling 95

Multiple Intelligences 96
Multivoting 97
Myers–Briggs Type Indicator 97

N 99

Negotiating 99
Net Present Value 100
Network Diagram 100
Neural Nets 102
Nominal Group Technique 103

O 105

Object and Process Models 105
Objectives 106
Organization Chart 107
Organizational Engineering 108
Organizing 109
Outsourcing 109

P 111

P^{2}M^2 111
Paradigm 111
Pareto Analysis Chart 112
Parkinson's Law 113
PDCA Cycle 114
Peak Experience 115
PERT Estimates 116
Peter Principle 117
Planning 118
Post-Implementation Review 118
Power 119
Presentation 120
Presentation: Perception 120
Presentation: Performance 121
Presentation: Perspective 121
Presentation: Planning 122
Presentation: Practice 122
Presentation: Preparation 123
Priorities of Change 123
Probability 124
Problem Solving 125
Procedures 126
Project 126
Project History Files 127

Project Library 128
Project Life Cycle 128
Project Management 129
Project Management Software 130
Project Manager 132
Project Manual 132
Project Newsletter 133
Project Office 134
Project Sponsor 134
Project Team 135
Project Wall 136
Project Website 137
Prototyping 137

Q 139

Quality Assurance 139
Quality 140

R 141

Reengineering 141
Regression and Correlation Analysis 141
Replanning 142
Reports 143
Requirements Definition 143
Resource Allocation 144
Responsibility Matrix 145
Reuse 146
Risk Analysis 147
Risk Assessment 147
Risk Control 148
Risk Management 148
Risk Reporting 149

S 151

Sampling 151
Scattergram 152
Scheduling 153
Scope Creep 154
Security 155
Self-Directed Work Teams 156
Senior Management 156
Six Hats 157
Skills Matrix 158
Social Behavior Typology 159

Span of Control 159
Speed Reading 160
Staff Meetings 161
Stakeholders 162
Standard Deviation, Variance, and Range 162
Statement of Work 164
Statistical Process Control 166
Statistics 167
Status Assessment 168
Status Collection 168
Status Review Meeting 169
Stewardship 170
Strategic Planning 171
Supplier Management 171
Supply Chain Management 172
Systems Theory 173

T 175

Team Building 175
Team Organization 175
Teaming Basics 176
Technology Transfer 177
Telecommuting 178
Testing 178
Time Management 179
Top-Down and Bottom-Up Thinking 180
Total Float 181
Tracking and Monitoring 182
Training 183
Transactional Analysis 183
Tuchman Model 184

U 187

Unity-of-Command Principle 187

V 189

Variance 189
Videoconferencing 189
Virtual Teaming 190

W 193

Winding Down 193
Work Breakdown Structure 193
Workflow Analysis 198

Work Flows 199
Workplace Design 200
Writing 200

References 203

ACTIVITY-BASED COSTING AND TRADITIONAL ACCOUNTING

Many companies employ traditional accounting practices that focus on costs. Most of these practices concentrate on removing overhead, reducing inventories, and allocating costs. The results are often an inaccurate portrayal of costs and a "hatchet" approach toward cutting those costs. Activity-based costing (ABC) takes a more direct approach towards cost accounting. It requires looking at costs from customer satisfaction and process perspectives. It enables determining which costs contribute to meaningful output and which ones are not effective or efficient. In other words, costs are associated with output, not a cost category *per se*. Project management plays a critical role in ABC by defining deliverables and tasks that execute specific processes on projects.

GOALS

- Facilitate decision-making.
- Focus on customer satisfaction.
- Determine product costs more accurately.
- Determine which improvements provide the most savings.
- Balance short- vs. long-term perspectives.

OBSTACLES

- Failure to recognize or appreciate the need for a paradigm shift
- Lack of patience or support when instituting ABC in a traditional functional accounting environment

STEPS

1. Recognize that a paradigm shift is required to move from a functional accounting perspective to an ABC perspective.
2. Understand the transition does not come easily.
3. Train people in the basics of ABC.
4. Apply ABC on a pilot scale.
5. Conduct an impact analysis on the transition from functional to ABC accounting.

AFFINITY DIAGRAM

An affinity diagram is a graphical technique for taking a collage of items and dividing it into groups based upon specific, similar characteristics (Figure 1). The diagram allows people to comprehend a myriad of items and draw some preliminary conclusions from the grouping of items.

GOALS

- Enable better management of data and information.
- Enable greater understandability of data and information.
- Communicate with other stakeholders about data and information.

OBSTACLES

- Using unclear criteria for grouping items
- Trying to "force" an item into an incorrect grouping

STEPS

1. Compile all the necessary data.
2. Determine the natural groupings or categories of data.
3. Hierarchically arrange categories.
4. Assign each datum to the appropriate category.
5. For datum not fitting within a category, form a separate category or place it in the closest category.

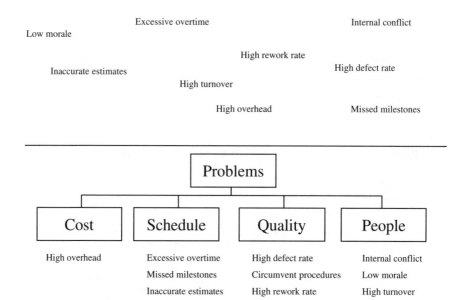

FIGURE 1 Affinity diagram.

ALTERNATIVE WORKING SCHEDULES

Alternative working schedules have become a way of life for many corporations. Some examples of alternative working schedules are flex-time and compressed work weeks. The idea is to provide some flexibility to employees in their schedules to reduce absenteeism and tardiness. Alternative working schedules require some good coordination, however, to realize the benefits. One coordination issue is to have a core set of hours available during a day and throughout a week whereby everyone on a team is available. This coordination is absolutely necessary for projects dealing with mission-critical systems. Another coordination issue is to have at least one person available on site at all times to ensure ongoing communications internally and externally to the project. Alternative working schedules require an element of trust by management to be administered effectively.

GOALS

- Reduce the job time (e.g., absenteeism).
- Increase morale.

OBSTACLES

- Not providing for "core availability" times
- Not providing the trust to allow alternative working schedules to succeed

STEPS

1. Identify who will participate on an alternative schedule.
2. Determine the core hours.
3. Ensure a point of contact exists for any periods when everyone is gone.

B

BAR (GANTT) CHARTS

A bar chart, also known as a Gantt chart, provides an easy-to-read view of the flow time, or duration, of a set of tasks at a particular level in a work breakdown structure (Figure 2). Frequently, the chart is drawn at a summary level and presented to the client and senior management because of its ease of understanding. A bar chart has, however, some shortcomings. It does not reflect relationships among the tasks, critical path, and early and late start and finish dates. At a minimum, a bar chart should contain these items: a work breakdown structure (e.g., deliverable or task listing); bars reflecting duration (colored in to reflect percent complete) and either early or late dates and atime scale to show flow times.

Goals

- Provide a path to reach a vision, goal, and objective.
- Maintain focus on a vision, goal, or objective.
- Enable better communications.
- Provide an effective reporting tool.

Task	Year	20XX		
	Month	April	May	June
	Week	1 2 3 4	1 2 3 4	1 2 3 4
Perform A		▬▬▬ 100%		
Perform B		▬ 25%		
Perform C			▭	
Perform D			▭	
Perform E			▭▭	
Perform F				▭
Perform G				▭

FIGURE 2 Bar chart.

OBSTACLES

- Using the chart for the wrong audience
- Crowding the chart with too much information as a way to compensate for shortcomings
- Not developing the chart in concert with a network diagram

STEPS

1. Identify the audience for the bar chart.
2. Use the work breakdown structure to identify the level reflected in the bar chart.
3. Reflect the responsibilities, flow time, and status for each task in the chart.
4. Coordinate contents of the chart with the supporting network diagram.

BENCHMARKING

Benchmarking is an approach for identifying processes and objects, for example, and comparing them to others to determine best practices. A standard or measure common to all of these processes or objects is used against which to benchmark. That standard or measure frequently involves a process or object of a premier company or project.

GOALS

- Identify a measure to determine a desired level of performance.
- Compare one's own performance against others.
- Provide a way to determine whether a need exists for improving performance.

OBSTACLES

- Using too vague a standard against which to compare processes
- Introducing bias into the benchmarking

STEPS

1. Determine the goal of the benchmarking.
2. Identify what (e.g., processes or practices) to benchmark.
3. Document the current process or practice.
4. Compare the current process or practice with that of other organizations using a common standard.
5. Identify any gaps.
6. Determine improvements.
7. Implement improvements.
8. Monitor performance.
9. Conduct benchmark once again, if necessary.

BODY LANGUAGE

It has been estimated that at least 70% of our communications is exhibited through body language. From a communications perspective, the delivery and reception of a message are influenced greatly by body language, such as the use of the arms, head, eyes, and legs. Body language also involves, for example, space, touch, and sound. The key point to remember is that body language, in order to prove useful, must be observed over a series of actions, over a period of time, and within context.

GOALS

- Pick up on actual messages being sent or filtered.
- Ensure clear delivery and reception of messages.

OBSTACLES

- Relying upon one type of body language (e.g., crossed arms)
- Observing body language over a short period of time
- Jumping to conclusions about certain signals
- Failing to recognize cultural differences in body language

STEPS

1. Avoid the tendency to rely on one body signal; look for patterns over a period of time.
2. Look at the context of the situation before jumping to conclusions.
3. Recognize that body language should be interpreted along with the speech (e.g., voice and word choice).
4. Recognize that cultural differences exist.
5. Use own body language to express a point, win concurrence, or solidify a relationship with stakeholders.
6. Remember that body language involves the entire body, not just selected parts.

BRAINSTORMING

Brainstorming is a free-flowing approach for generating a list of ideas, options, etc. It involves having a group of people, preferably no more than ten, generating a random set of ideas related to a particular topic or solution to a problem. The session is free and open, meaning people suspend judgment. Every idea is written on a white board, for instance, without regard to quality; the emphasis is on quantity. After developing all the ideas, evaluation can occur. An interesting technique used in brainstorming is called *hitchhiking*, which involves piggybacking off one idea to generate another idea and so on. The typical sequence for a brainstorming session is problem or issue definition, brainstorming, and then evaluation by eliminating, combining, and refining ideas.

Goals

- Generate as many ideas as possible.
- Avoid judgmental thinking.

Obstacles

- Allowing laughter and ridiculing of ideas
- Emphasizing quality over quantity

Steps

1. Follow the sequence of problem or issue definition → brainstorm → evaluation.
2. Define the problem or issue to address.
3. Assemble all the selected participants in a room with comfortable surroundings.
4. Try to have a diverse group of participants.
5. Remind people to forego any explicit or tacit evaluation of ideas until later.
6. Have participants call out ideas to be put on an easel pad or white board.
7. Apply the concept of hitchhiking or ping pong of ideas.

BREAKEVEN ANALYSIS

Breakeven analysis is also known as the payback period method (Figure 3). Its purpose is to determine when a proposed project will pay for itself in a specific

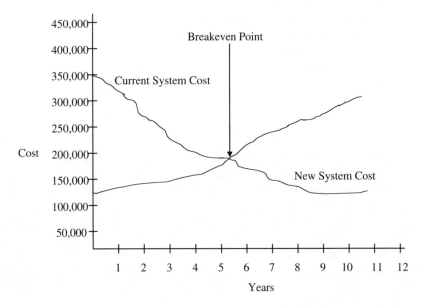

FIGURE 3 Breakeven analysis.

number of years *vis-à-vis* staying with the *status quo*. If multiple projects are considered, then the one with the shortest payback period is selected. The key variables for calculating the payback period are the number of years, the cost of maintaining the *status quo*, the cost of the proposed project, cumulative years, and the corporate tax rate, if applicable. The before-corporate tax rate formula is:

Breakeven point = cost of proposed project ÷ average return on investment

For example,

$$5 \text{ years} = \$50,000 \div \$10,000$$

The after-corporate tax rate is:

Breakeven point = cost of proposed project ÷ [(1 − tax rate)(average return on investment)]

For example.

$$6.7 \text{ years} = \$50,000 \div [(1 - .25)\$10,000]$$

GOALS

- Determine if a proposed solution will pay itself back over a shorter period of time vs. staying with the *status quo*.
- Weigh short- and long-term costs and perspectives.

OBSTACLES

- Failing to recognize that breakeven analysis is only one method of analysis
- Using inaccurate or exaggerated data in variables

STEPS

1. Identify the following for each scenario:
   a. Investment
   b. Average return on investment
   c. Corporate tax rate
2. Calculate the payback period before and after taxes, if applicable, for each project.
3. Plot the results of the calculations.
4. Select the desired option.

BUDGETING

Budgeting involves allocating funds to plan, organize, control, and lead a project. It occurs once all the cost estimates have been developed. More often than not, labor is the biggest budget item because it incurs the biggest costs; however, this scenario is not always the case. It makes good sense, therefore, to remember that a budget should account for other items such as time penalties, information acquisition, equipment usage, training, traveling, and facilities. It is also important to note that a budget should account for all the different types of costs that can occur on a project, such as:

- *Direct costs*: costs directly involved in making or assembling a product or delivering a service vs. indirect costs (all costs other than direct costs, such as rent, taxes, insurance)
- *Recurring costs*: costs that appear regularly vs. nonrecurring costs (costs that occur once, such as equipment purchase)
- *Fixed costs*: costs that stay the same as work volume changes vs. variable costs (costs that vary depending upon consumption and workload)
- *Burdened rate*: cost of fringe benefits (e.g., insurance and floor space) and overhead vs. nonburdened rate (excludes costs of fringe benefits and overhead)
- *Regular rate*: less than or equal to 40 hours per week vs. overtime rate of greater than 40 hours per week, which includes time and a half and double time

Goals

- Provide money for leaner times.
- Increase a company's profit margin.
- Provide more funds for critical activities.
- Price a product or service competitively.

Obstacles

- Budgeting too much or too little
- Budgeting according to wants rather than needs
- Overlooking certain categories of costs

Steps

1. Define roles and responsibilities for preparing, reviewing, and approving budget.
2. Define acceptable, calculated rate for specific resources.
3. Develop a process or procedure and reports for reporting performance against budget.
4. Tie budgeting to the work breakdown structure, time estimates, and schedule developed for the project.

C

CAPABILITY MATURITY MODEL

The capability maturity model (CMM), by the Software Engineering Institute (SEI), is a model for determining the level of maturity in a company's software engineering practices. CMM consists of five levels, each one divided into key process areas and further subdivided into key practices. The idea is that all organizations move from the first to the fifth level; it is rare that an organization goes beyond level 3. The *initial* maturity level (level 1) is one where software engineering is chaotic and *ad hoc*. The *repeatable* level (level 2) is one where project management, requirements management, and configuration management, for example, are in place. The *defined* level (level 3) is one where the focus is on processes, particularly their integration, coordination, and documentation. The *managed* level (level 4) is one where quantitative measures are in place for process and product performance. The *optimizing* level (level 5) is one where defect prevention and innovation become commonplace. The approach for determining the level of an organization or project is through an assessment.

GOALS

- Identify the level of maturity through assessment.
- Institute processes and practices that lead to a higher level of maturity.

OBSTACLES

- Allowing CMM to become a bureaucratic endeavor, thereby slowing project performance
- Not conducting an objective assessment
- Not taking the time or effort to understand concepts behind CMM

STEPS

1. Obtain a good knowledge and understanding of SEI's CMM.
2. Recognize the need for conducting an objective assessment.
3. Avoid the tendency to allow CMM to weigh down project performance.
4. Recognize that very few organizations achieve level 3 (defined level) and above.
5. Recognize that CMM is designed to help a project achieve its objectives, not make compliance the objective.

CAPACITY PLANNING

Capacity planning is determining the current capabilities and anticipating the future needs of a customer regarding some delivery system, such as a local area network or production line. A good capacity plan takes into consideration historical information, priorities, scheduling, metrics, throughput, processes (e.g., business functions), and critical success drivers (e.g., business and technical drivers), patterns of behavior (e.g., usage), and interactions among elements. The three overall criteria for assessment of existing and future needs are the current and desired levels of efficiency, effectiveness, and utilization. It is important to remember that capacity planning is an ongoing endeavor because business environments change constantly.

GOALS

- Apply existing resources efficiently.
- Anticipate future needs of resources.
- Provide the ability to adapt to changing demands for resources.

OBSTACLES

- Unavailability of data
- Failure to document existing infrastructure
- Failure to maintain configuration control
- Failure to identify all major priorities and variables

STEPS

1. Identify key business processes and critical success factors.
2. Identify technical and business constraints.
3. Obtain historical data.
4. Determine criteria for measuring capacity from efficiency and effectiveness perspectives.
5. Take any measurements to address shortfalls in historical data.
6. Balance current with future requirements.
7. Document the capacity.
8. Periodically revisit the plan.

CATEGORIES OF CHANGE

Changes can impact schedule, cost, and quality; however, not all changes are of equal magnitude. Some changes are classified as "major" changes, which alter one of three items dramatically. Some changes are "minor" changes, which alter one of the three items but less so than a major one. Minor changes do not alter the outcome of a project in a big way. Some changes are "corrective" changes, which address something that was very insignificant and originally overlooked.

GOALS

- Provide better management of incoming changes.
- Enable better allocation of resources.

OBSTACLES

- Inability to define each category
- Failure to define and follow response to a particular category of change

STEPS

1. Develop medium (e.g., form) for capturing information regarding a change.
2. Develop criteria for categorizing a change.
3. Apply criteria.
4. Notify the person requesting the change, if necessary.

CAUSE-AND-EFFECT GRAPH

A cause-and-effect graph is a graphical way to show the relationships between one or more causes and one or more corresponding results (Figure 4). The typical symbols are a circle or node to represent a cause and a vector to represent a path

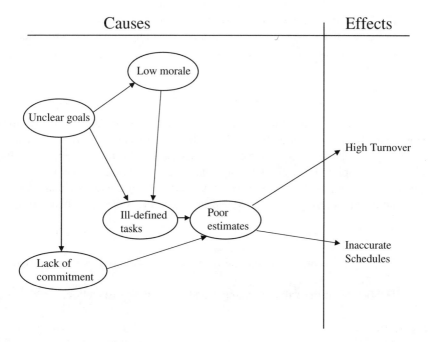

FIGURE 4 Cause-and-effect graph.

leading to a specific result. Such a graph is used to understand the true cause or causes of a particular result.

GOALS

- Identify relationships between causes and effects.
- Communicate easily the relationships between causes and effects.
- Clarify relationships between causes and effects.

OBSTACLES

- Failure to identify all the relevant causes and effects
- Failure to identify all the relationships between causes and effects
- Failure to keep strings short and manageable
- Failure to assume correctly about the relationships between causes and effects

STEPS

1. Determine:
   a. Purpose of the graph
   b. Audience
2. Identify all causes and effects.
3. In some cases, identify all intermediate causes.
4. List all the effects on the right side of a page.
5. From left to right, write the nodes representing causes on the paper.
6. From the effects, work from right to left connecting all the nodes to reflect the chain of causes resulting to the effects.

CHANGE BOARD

A change board is a group of individuals who meet regularly to classify and prioritize changes and assess whether a change should be implemented. Specific responsibilities include: categorizing incoming changes; assigning priorities to them; analyzing their impact; approving or disapproving them; and assigning responsibilities for implementing changes. A typical change board consists of a project manager, team leads, and client representatives. Most decisions are determined by a majority vote.

GOALS

- Obtain involvement and commitment of key stakeholders to requested changes.
- Deliberate on the value of changes and direct their disposition.

OBSTACLES

- Failure to identify the roles and responsibilities of stakeholders.
- Failure to have the participation of key stakeholders.

Change Control

STEPS

1. Identify stakeholders affected directly by changes.
2. Develop process or procedure for identifying and assessing changes via change board.
3. Conduct regularly scheduled change board meetings.
4. Document results of meetings.
5. At each meeting, cover upcoming changes as well as decisions regarding changes that have had a technical, schedule, and cost analysis.

CHANGE CONTROL

Change control consists of policies and procedures established to detect, analyze, evaluate, and implement modifications to all baselines in a project. Changes can occur to schedule, budget, and quality criteria. A baseline, of course, is an agreement between two or more parties on what constitutes something, such as product, description, schedule, and budget. Changes can come from many sources, such as the result of reviewing status, requests or demands from stakeholders, and mandates from external authorities (e.g., a governmental body). They can have far-reaching impacts on how stakeholders feel about a project, work processes, schedules, budgets, and quality criteria. The key is to manage change and not be managed by it. A change board and change prioritization and categorization can help manage change.

GOALS

- Reduce the possibility of scope creep.
- Evaluate changes to baselines.
- Obtain buy-in from stakeholders for changes.

OBSTACLES

- Failure to categorize changes
- Failure to prioritize changes
- Failure to assess the impact of changes on processes and products
- Failure to document changes
- Failure to establish baselines

STEPS

1. Establish baselines for product and schedule.
2. Develop criteria for classifying proposed changes.
3. Conduct a technical, schedule, and cost analysis for each proposed change.
4. Assess the impact of proposed changes.
5. Determine the disposition of proposed changes.
6. For accepted changes, determine their release schedule.
7. For a change that is not accepted, notify the person who made the request.

CHANGE IMPLEMENTATION

From a project management perspective, change does not come out of the barrel of a gun. It requires a more subtle approach that takes time and patience, particularly if the change is to be long lasting. Several key ingredients are necessary to implement change effectively. A vision must exist that all key stakeholders can embrace. Likewise, the same is true for the plan. Commitment and buy-in are important factors throughout implementation of a major change. The current climate, history, and culture of an organization are significant ingredients in the acceptance or resistance to a major change. The impact of processes, people, and technology must also be considered. In the end, effective change implementation requires recognizing the impacts that accompany it and providing the means to allow an organization to adjust to a change. Projects, of course, are harbingers of change because they develop a new product or deliver a new service that will challenge the *status quo*.

GOALS

- Implement a change effectively and efficiently.
- Encourage commitment, trust, and buy-in.
- Reduce or eliminate fear.
- Posture a change as a win–win scenario.

OBSTACLES

- Overlooking the people side of change
- Failing to identify impacts of a change
- Treating change as a win–lose scenario

STEPS

1. Determine the goals and objectives of the change.
2. Develop an overall strategy and selected tactics for implementing change.
3. Incorporate goals, objectives, strategy, and tactics in a plan.
4. Recognize that change not only affects processes, practices, and profits but people, too.
5. From a people perspective, consider factors such as culture, recognition, buy-in, trust, feedback, commitment, management support, and history.
6. Tolerate failure.
7. Identify external and internal forces positively and negatively affecting the implementation of a change.

CHECKPOINT REVIEW MEETING

A checkpoint review meeting is held at the end of a phase or the completion of a major milestone. The purposes of the meeting are to determine the adequacy of the work completed and whether or not to proceed or to perform rework. Stakeholders,

Chunking

such as the project manager, customer representatives, sponsors, and steering committee members, often attend the meeting. Minutes are taken and distributed.

GOALS

- Determine whether or not to proceed from one phase or milestone to another.
- Check whether or not a project will deliver results as expected.
- Communicate important information to stakeholders.

OBSTACLES

- Failing to follow an agenda
- Failing to take minutes
- Not having the necessary stakeholders present
- Failing to make key decisions
- Allowing the meet to be dominated by a few people

STEPS

1. Prepare an agenda.
2. Maintain focus on it.
3. Invite the necessary stakeholders.
4. Ensure that the proper location, replete with supplies, equipment, etc., is available.
5. Ensure that everyone participates during the session, if appropriate.
6. Concentrate on facts and data.
7. Take and publish minutes.
8. Schedule (e.g., at the end of a phase), when necessary.

CHUNKING

Chunking is the process of dividing a large item, such as a product or concept, into smaller parts based upon some criteria (Figure 5), by identifying natural divisions that might exist. Through chunking, a person can better understand the product or concept.

GOALS

- Improve comprehension of data and information.
- Improve communication of data and information.

OBSTACLES

- Not taking advantage of the natural divisions among components
- Not dividing a process or object into a sufficient number of small components
- Dividing a process or object into too many unmanageable, small components

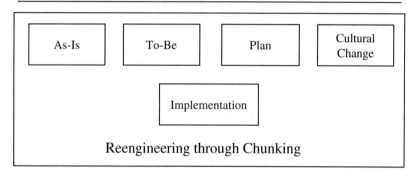

FIGURE 5 Chunking.

STEPS

1. Determine:
   a. Goal of chunking
   b. Audience
2. Identify the natural divisions of the item being subdivided and incorporate them in the criteria for chunking.
3. Determine the extent of chunking.
4. Apply chunking.
5. Refine the results until the goal has been achieved.

CLIENT

The client is the person or group for whom the product is being built or the service is being delivered. Specific responsibilities include receiving deliverables, approving the final deliverable, communicating requirements, coordinating with other stakeholders, and providing dedicated resources.

GOALS

- Provide the necessary resources, such as people, with the requisite skills, education, and experience.
- Provide a clear set of requirements and specifications.

OBSTACLES

- Not providing clear requirements and specifications
- Not following through on responsibilities
- Allowing internal bickering over requirements and specifications

STEPS

1. Determine formal and informal roles and responsibilities.
2. Determine the context of the environment.
3. Determine formal and informal powers.
4. Determine role expectations.
5. Determine levels of commitment and support.

COLLECTING STATISTICS

Collecting statistics is done for costing, scheduling, and quality control purposes. For collecting statistics to be useful, it must be performed regularly and consistently, using reliable, valid approaches that reduce biases to give an objective picture of project performance.

GOALS

- Obtain data on the performance of a project for audit purposes.
- Obtain ideas for improving opportunities and capitalizing on them in the future.
- Enable thorough analysis of performance of a project.
- Improve management of similar projects in the future.
- Develop databases for cost and time estimates.

OBSTACLES

- Entering bias into the collection process
- Collecting statistics for statistics' sake
- Not using a consistent, reliable approach toward collecting statistics

STEPS

1. Develop a consistent and reliable process or procedure for collecting and compiling data and generating information.
2. Determine the audience for data and information.
3. Determine the eventual format of information.
4. Determine the frequency and method for collecting data.

COMMUNICATION DIAGRAM

A communication diagram is way to depict interactions among different stakeholders on a project (Figure 6). It provides a key understanding of how people interact and to what extent. The diagram serves as an excellent vehicle, for example, to determine who communicates what messages to whom and to what degree. This information can prove very important when considering workplace design modifications or changing ways of performing work. It also can help in distinguishing between the formal and informal power structure in an organization or on a large project.

GOALS

- Identify all the key stakeholders.
- Identify the most prevalent or most important relationships.
- Distinguish between the formal and informal power structure.

OBSTACLES

- Failing to take the time or make the effort to adequately track communication relationships among stakeholders
- Failure to define what constitutes communications (e.g., data, paper) among stakeholders

STEPS

1. Identify stakeholders.
2. Determine the medium (e.g., message exchanges) for ascertaining interaction among stakeholders.
3. Determine the quality of those relationships.
4. Using objective criteria, determine the best arrangements or approaches to improve relationships among stakeholders.

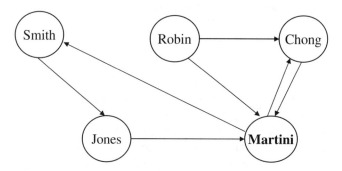

FIGURE 6 Communication diagram.

COMMUNICATIONS PROCESS

Communications on projects can be extremely complex among major stakeholders (client, project manager, project sponsor, and team). Each operates from a different perspective in terms of knowledge, skill, beliefs, and expectations. Their perspectives can result in a breakdown in communications. They filter incoming and outgoing messages for each stakeholder. The result is a host of misunderstanding and miscommunications. Failure to recognize this fact can wreak havoc on the efficiency and effectiveness of all stakeholders, especially the team.

GOALS

- Ensure greater cooperation.
- Facilitate better understanding.
- Improve overall teaming.

OBSTACLES

- Using too much jargon
- Going through too many layers of management
- Using stakeholders with varied backgrounds and experiences
- Using people with different beliefs and value systems
- Appling poor listening skills
- Failing to exercise empathy

STEPS

1. Determine stakeholders in the communications process of a project.
2. Identify common media.
3. Identify factors affecting quantity and quality of communications.
4. Determine opportunities for improvement.
5. Schedule improvements.
6. Establish feedback mechanisms on the effectiveness of improvements.

CONFIGURATION MANAGEMENT

Configuration management is an administrative function responsible for maintaining the overall integrity of a product and any related deliverables produced during and beyond the project life cycle. It consists of four functions. Identification is documenting the characteristics of a product and its associated deliverables with the purpose of establishing a baseline and instituting version control. Audit is ensuring conformance to requirements and specifications. Change control is managing and documenting changes to a product and its deliverables. Status accounting is recording and reporting of proposed and accepted changes.

GOALS

- Manage changes against a baseline.
- Maintain the integrity of project deliverables and final product.
- Provide an audit trail.
- Ensure conformance to specifications.

OBSTACLES

- Not establishing a configuration baseline for all applicable items on a project
- Not controling changes to any established baselines

STEPS

1. Identify all baselines related to cost, schedule, and quality.
2. Establish procedures for maintaining baseline integrity.
3. Assign someone responsible for configuration management responsibilities.
4. Conduct business and technical impact analyses of all significant changes.

CONFLICT RESOLUTION

Perhaps the most difficult aspect of managing or participating on projects is dealing with conflict. It can arise over a host of factors, such as differences in opinion as well as over the use of scarce resources. Of course, conflict by itself is not necessarily negative. It depends on how it is managed. The key is to focus on the problem or issue and not the people and their personalities. That means defining the problem, concentrating on the facts and data, and seeking a win–win outcome, if possible. All this requires everyone, particularly the project manager, to keep cool, relate (not berate), and focus on the goals and objectives of a project. It is imperative, too, that project managers are aware of their own signals as well as those of others involved in a conflict.

GOALS

- Motivate people to work efficiently and effectively.
- Increase morale and *esprit de corps*.
- Augment teaming.

OBSTACLES

- Not seeking a win–win solution
- Chastising or berating a team member
- Inability to pull away from the emotions of the situation and focus on the facts and data
- Failing to follow-up on the solution to the conflict

STEPS

1. Define the problem or issue.
2. Determine all the interests of the parties.
3. Focus on facts and data, not people and their emotions.
4. Keep the vision of the project in the center of everyone's mind.
5. Observe the win–win result, if possible.
6. Encourage people to come up with a solution.
7. Perform follow-up on the effectiveness of a solution.

CONSULTANTS AND CONTRACTORS

On any given project, consultants and contractors play key roles in the outcome. They often provide key skills and knowledge that other members on a team may not possess. Because they are temporary workers, it is important to ensure that they are not treated as employees, either by accident or design. Their treatment has implications for Internal Revenue Service (IRS) rules and regulations and provisioning of benefits. In addition, it is important to define their roles and responsibilities because the relationship with employees can be one of mistrust and jealousy, thereby affecting morale and, ultimately, productivity.

GOALS

- Obtain qualified people when necessary during the life cycle of a project.
- Avoid violation of IRS rules distinguishing consultants and contractors from employees.
- Provide a diversified team in regard to expertise and knowledge.

OBSTACLES

- Monitoring quality of work of short-term participants
- Creating rivalries among consultants, contractors, and employees
- Avoiding a quick-fix mentality

STEPS

1. Refer to the rules and regulations of the Internal Revenue Service on contractors, consultants, and employees.
2. Whether for contractors or consultants, consider performing these actions:
   a. Conduct interviews.
   b. Conduct background investigations.
   c. Obtain references.
   d. Inspect previous work.
   e. Deal with them in writing (e.g., contractually).
   f. Include their tasks in the work breakdown structure.
3. Give preference to employees over contractors and consultants.

4. Understand the potential impact to a project if senior management decides to let go consultants and contractors.
5. Determine the risks associated with using consultants and contractors.

CONTINGENCY PLANNING

Contingency planning is anticipating responses to circumstances that can negatively impact a project. Often, such circumstances deal with cost, schedule, quality, and people. Such planning requires determining the necessary steps to overcome problems occurring on a project. It should, of course, be closely coordinated with risk-management activities on a project.

GOALS

- Minimize negative occurrences on a project.
- Provide for flexibility in response to changing situations.
- Instill a sense of confidence in stakeholders.
- Maintain the momentum of a project in the face of anticipated obstacles.

OBSTACLES

- Inability to recognize the major risks of a project
- Inability to determine an appropriate response to a given situation
- Failure to document contingency plans
- Failure to rehearse plans

STEPS

1. Identify potential obstacles threatening the scheduled performance, determine their negative impacts, and determine an appropriate response.
2. Identify potential obstacles relating to cost performance, determine their negative impacts, and determine an appropriate response.
3. Identify potential obstacles relating to quality performance, determine their negative impacts, and determine an appropriate response.
4. Identify potential obstacles relating to employee performance, determine their negative impacts, and determine an appropriate response.
5. Document the potential obstacles, their negative impacts, and planned response.

CONTINUOUS IMPROVEMENT

Continuous improvement (or, as it is commonly known, *kaizen*) is improving an existing process incrementally. The idea is that a process, through measurement and monitoring, can be improved toward a level of perfection. It uses a wide range of tools and techniques, such as a fishbone diagram and statistical process control.

Goals

- Improve the efficiency and effectiveness of processes.
- Establish means for feedback on performance.

Obstacles

- Failing to follow-up on recommendations for improvement
- Using the wrong or insignificant measurements
- Using inaccurate, unreliable data for calculations
- Conducting too much refinement, resulting in diminishing returns

Steps

1. Determine:
   a. Goal to achieve
   b. Tool or technique to apply
2. Document existing process.
3. Collect data on existing process.
4. Conduct measurement.
5. Determine change to implement.
6. Implement change.
7. Conduct measurement.
8. Determine if improvement has occurred.
9. Make any necessary revisions.

CONTRACTS

A project is often the result of a contractual agreement between companies. The agreement can be a firm, fixed-price type, where a firm provides a service or builds a product for a fixed price. The provider or builder assumes a much higher risk than the buyer. Or, the agreement can be one where the buyer provides the supplies and overhead, and the provider or builder receives a certain level of profit. In this case, the buyer assumes most of the risk. Between these two vast types of agreements lies a series of derivative ones, such as cost plus fixed fee, fixed price with escalation, cost plus incentive fee, and cost without fee, to name a few. Regardless of the type of agreement, it is important to define clearly the terms and conditions, roles and responsibilities, requirements, payment schedules, cost and schedule targets, and exit clauses.

Goals

- Have the right type of contract for the circumstances.
- Have clear terms and conditions in the contract.
- Engage in a mutually beneficial contract.

OBSTACLES

- Unclear terms and conditions contained in contracts
- One-sided relationships that provide no exit or opportunity for redress

STEPS

1. Determine the type of contract desired.
2. Determine the terms and conditions.
3. Ensure that the terms and conditions are clearly defined.
4. Ensure that the terms and conditions cover these areas:
   a. Fees
   b. Risks
   c. Costs
   d. Roles
   e. Responsibilities
   f. Requirements
   g. Deliverables
   h. Payment schedule
5. Consider also the general climate (e.g., tense, prosperous) surrounding the negotiation of a contract.
6. Encourage discussion of assumptions and expectations on the part of all parties.

CONTROLLING

Controlling is one of the four functions of project management. This function assesses how well the plans and organization are being utilized to meet the visions, goals, and objectives of projects. It includes tracking and monitoring, collecting and assessing status, planning for contingencies, responding to critical problems, conducting effective meetings, and closing projects efficiently and effectively. Many projects reflect one of three scenarios: management by confusion, management by drive, and management by direction. Management by confusion occurs when there is no sense of direction for a project. Management by drive occurs when nothing happens for a long period of time and then a mad rush occurs to complete a project at the end of its life cycle. Ideally, projects should seek management by direction, where productivity is constant and oriented toward achieving the vision, goals, and objectives.

GOALS

- Maintain focus on vision, goals, and objectives.
- Apply resources efficiently and effectively.
- Obtain valid, reliable feedback on progress.
- Integrate with the other functions of project management.

Obstacles

- Failing to recognize that a project tends to move toward anarchy
- Collecting and relying upon inaccurate feedback
- Yielding to the 90% syndrome (considering the project 90% completed, with the remaining 10% lasting indefinitely) to collect status
- Failing to recognize the importance of Parkinson's law
- Failing to allow decisiveness by relevant stakeholders

Steps

1. Develop a process for collecting data and information on feedback related to project performance.
2. Document the process (e.g., prepare a procedure).
3. Publish the process (e.g., place in a project manual).
4. Follow the process (e.g., conduct ongoing tracking and monitoring).

CONTROLS

Controls are measures in place to mitigate or eliminate the impact of actual risks facing a project. A control can be in place for a process, procedure, or component. Basically, three types of controls exist: preventive, detective, or corrective. Preventive controls are in place to ensure that occurrence of a potential risk has no impact on a project. Detective controls identify the impact of a potential risk should it affect a project. Corrective controls are implemented to reduce or eliminate the effects of a risk that has become reality.

Goals

- Reduce or eliminate the impact of a risk.
- Reduce the amount of oversight.
- Enable focusing on the goal or goals.

Obstacles

- Applying the wrong control to address a risk
- Assuming that a control will be effective without testing or prior knowledge
- Failing to review the applicability or effectiveness of a control after a period of time

Steps

1. Identify the risks that occur on the project.
2. Identify the major processes or components.
3. Determine the controls that do or should exist.
4. For existing controls, ascertain their effectiveness.

5. For processes or components deemed unprotected, determine and implement the appropriate controls.

CORE TEAM

A core team consists of those stakeholders who contribute the essential skills, knowledge, and talents to the development of a product or delivery of a service (Figure 7). As an exercise, project managers can determine their core teams by considering the likelihood of forced reductions in the team memberships and determining whom they would retain and release. The core team would, of course, be the people chosen to remain.

GOALS

- Provide a reliable group of individuals throughout the life cycle of a project.
- Increase consistency, efficiency, and effectiveness of performance.

OBSTACLES

- Inability to designate which team members should be on the core team
- Assign core team members to noncritical tasks while non-core team members work on critical tasks
- Not have backups for core team members

STEPS

1. Identify all the stakeholders on the project team.
2. Determine the goals, objectives, and tasks.

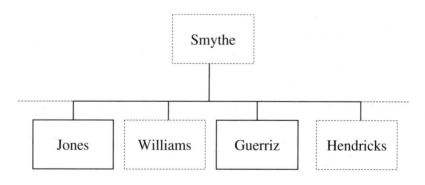

FIGURE 7 Core team.

3. Conduct an inventory of essential skills, knowledge, and experience.
4. Map the skills, knowledge, and experience of team members to the goals, objectives, and tasks.
5. Identify the people who contribute directly to the goal, objectives, and tasks.

COST ANALYSIS

Cost analysis is useful for evaluating investment, for example, in new equipment or a project. The keys to solid cost analysis are completeness and accuracy. Some major considerations include short- and long-term costs, borrowing costs (e.g., interest rates), direct and indirect costs, the time value of money, the reliability of data used in calculations, external costs (e.g., taxes), and assumptions. Credibility is very important and best accomplished through an objective review of the analysis (e.g., peer reviews).

GOALS

- Balance short- and long-term interests.
- Determine a reliable, valid cost estimate to make decisions regarding costs.
- Identify the key independent and dependent variables in the cost equation.

OBSTACLES

- Failing to consider all the key variables
- Overlooking the time value of money
- Inability to identify the impact of subtle, extraneous events

STEPS

1. Identify the key independent and dependent variables.
2. Collect the appropriate data to perform the cost analysis.
3. Strive for objectivity in calculations.
4. Account for the time value of money and the cost of borrowing money, if applicable.
5. Seek an objective review of all calculations.

CREATIVITY

Creativity is the capacity to generate novel, original ideas. It involves developing truly different ideas and techniques that build upon or replace prevailing business practices. Creative individuals have the ability to look beyond routine patterns of thinking and doing, thereby developing new insights that challenge the *status quo* in a given field. To a large extent, creativity involves coming up with breakthrough rather than incremental thoughts, tools, techniques, etc.

GOALS

- Develop innovative ideas, approaches, etc. for solving problems and overcoming obstacles.
- Augment *esprit de corps*.
- Encourage emotional commitment by stakeholders.

OBSTACLES

- Not providing resources conducive to being creative
- Having a low tolerance for failure
- Failing to reward successful creative behavior and results

STEPS

1. Determine the appropriate time and place for being creative.
2. Allow for failure as much as for success.
3. Recognize, even if failure occurs, the effort made to develop and implement something creative.
4. Provide resources for being creative.
5. Encourage risk taking.
6. Reward creativity as much as being right.

CRITICAL CHAIN

Critical chain, developed by Eliyahu Goldratt, is the theory that the longest path in a project schedule consists of the combination of task dependencies and resource capacities. The successful implementation of tasks depends on their timely completion and resource availability, particularly in the case of sequential, rather than concurrent, tasks. A related concept is the importance of adding buffers to tasks and, if necessary, for an entire project. These buffers adjust for the exaggeration and understatements that estimates often incorporate. For example, a person can use an "average" estimate and adjust it by a probability factor of 50%.

GOALS

- Recognize significant constraints on a project.
- Reduce the impact of unrealistic estimates.

OBSTACLES

- No clear understanding of critical chain concepts
- Overemphasizing the importance of resource constraints and buffers

STEPS

1. Determine critical estimates to complete tasks.
2. Identify opportunities to incorporate buffers.

3. Identify resource requirements and their accompanying constraints.
4. Establish important trigger points in schedules.

CRITICAL ISSUES AND ACTION ITEMS LOG

The critical issues and action items log (Figure 8) is a medium for collecting data on the assignment of tasks that are usually not captured in the schedule for a project. Frequently, it is used at a status or checkpoint review meeting to record assignments. A typical log contains some or all of this information for each assignment: issue number, description, assignee name, assignment date, proposed resolution date, and additional comments. At subsequent status or checkpoint review meetings, reference is made to outstanding items listed in the log.

GOALS

- Engender accountability and responsibility for results.
- Maintain visibility of important issues or topics.
- Provide an audit trail.

OBSTACLES

- No follow-up on the contents of the log
- Omission of key critical issues or action items in the log

STEPS

1. Determine a scheduled time for capturing critical issues and action items.
2. Capture all the applicable data for each item.

Issue Identifier	Description	Assigned Date	Person	Due Date	Additional Remarks
1	Purchase 5 laptops	5/1/XX	Jones	7/15/XX	Emphasize functionality over price

FIGURE 8 Critical issues and action items log.

3. Give visibility to the log to encourage follow-through.
4. Follow up on the contents of the log.

CRITICAL PATH

The critical path contains those tasks in the schedule that cannot slide at any moment in time; otherwise, the project end date will not be met. The critical path is identified in a network diagram according to these criteria: (1) the path is the longest in terms of cumulative duration in a network diagram, (2) the early and late start and finish dates match, and (3) all activities that have the lowest float are on the path. Here are some useful insights to remember when working with the critical path: Always give priority to tasks on the critical path. The lower the float, the more critical the task. Multiple critical paths can exist in a schedule.

GOALS

- Identify the most important tasks.
- Employ resources efficiently and effectively.
- Finish the most important tasks on time and, consequently, an entire project.

OBSTACLES

- Not recognizing that the critical path changes constantly
- Not following or using a critical path

STEPS

1. Identify all the dependencies.
2. Calculate the total float for each task by performing forward and backward passes.
3. Identify tasks with minimum float (the longest path or paths through the network diagram).

D

DECISION MAKING

Decision making is being able to decide the appropriate course of action that will help further the attainment of vision, goals, and objectives of a project. It is a process that involves these activities: establishing goals and objectives, measuring their achievement, identifying problems, developing alternative decisions, selecting an alternative, implementing the selection, and controlling and evaluating progress. A common approach to implementing this process is through the use of the PDCA (plan, do, check, act) cycle (more commonly known as the Deming wheel).

Goals:

- Be decisive without being hasty.
- Further the achievement of goals and objectives.
- Employ resources efficiently and effectively.
- Balance short- and long-term requirements.

Obstacles

- Being indecisive or too hasty to make a decision
- Taking a quick-fix perspective
- Not considering the most important elements of information
- Not obtaining commitment to a decision
- Deviating from the vision, goals, and objectives

Steps

1. Determine the goals and objectives of a project.
2. Develop measures to determine their achievement.
3. Identify any potential problems, constraints, or risks.
4. Develop alternatives.
5. Select an alternative.
6. Implement the alternative.
7. Measure results attained.
8. Take corrective action, if necessary.

DECISION TABLES

A decision table is a matrix-like approach to showing the relationships between specific conditions and actions (Figure 9). It contains a condition column and an action column. A symbol placed in the cell where a condition and action intersect indicates the value of that relationship.

GOALS

- Identify relationships between conditions and actions.
- Communicate relationships between conditions and actions.
- Clarify relationships between conditions and actions.

OBSTACLES

- Failing to identify all the relevant conditions and actions
- Failing to identify all the relationships between conditions and actions
- Making incorrect assumptions about the relationships between conditions and actions

		Actions	
Conditions	Outsource	Renegotiate	Cancel Project
Poor schedule performance			★
Unsatisfactory product quality	★		
Exceed budget		★	
Low morale	★		
Legal problems			★
No available expertise	★		
High turnover	★		
Slow learning curve	★		
No management support			★

 Indicates action to take

FIGURE 9 Decision table. (From Project Management Seminar presented by Practical Creative Solutions, Inc., 1996.)

STEPS

1. Determine:
   a. Purpose of the decision table
   b. Audience
2. Draw a table, creating a condition stub (*y*-axis), an action stub (*x*-axis), and lines to create cells reflecting the relationship between a given set of conditions and actions.
3. Determine a symbol to reflect the value of any given relationship between conditions and actions.
4. Complete the decision table with title, subtitles, and legend.

DECISION TREES

A decision tree is a graphical approach for showing the relationships between specific conditions and actions (Figure 10). It consists of a series of lines representing a set of available choices made from one condition to the next.

GOALS

- Identify relationships among conditions and actions.
- Easily communicate relationships among conditions and actions.
- Clarify the relationships among conditions and actions.

OBSTACLES

- Failure to identify all the relevant conditions and actions
- Failure to identify all the relationships between conditions and actions

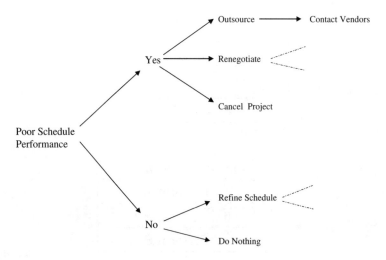

FIGURE 10 Decision tree.

- Strings become so long as to be unmanageable
- Failure to assume correctly about the relationships between conditions and actions

STEPS

1. Determine:
   a. Purpose of the decision tree
   b. Audience
2. Identify all conditions and actions.
3. From left to right (first condition, then actions) and from right to left (last actions to preceding condition), draw a decision tree.
4. Ensure that all conditions are connected to a given set or to more actions to take.
5. Ensure that no gaps exist in a decision tree.

DEPENDENCY RELATIONSHIPS

Dependency relationships are the associations among two or more tasks (Figure 11). The three basic types of relationships are finish-to-start (FS), start-to-start (SS), and finish-to-finish (FF). The FS relationship occurs when one task ends and the next task begins; SS occurs when one task starts and the next task starts without waiting

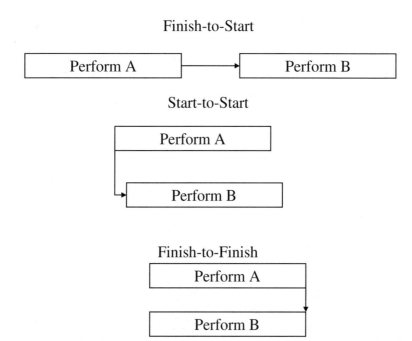

FIGURE 11 Dependency relationships. (From Project Management Seminar presented by Practical Creative Solutions, Inc., 1996.)

for the other one to finish; and FF occurs when two or more tasks must finish at the same time before subsequent tasks can begin.

GOALS

- Determine the type of dependency relations among tasks.
- Calculate the start and finish dates for each task and an entire project.
- Use these relationships to determine the critical path.

OBSTACLES

- Assuming that all tasks in a schedule have one type of dependency relationship
- Failing to account for lags in dependency relationships among two or more tasks

STEPS

1. Determine the logical sequence of tasks.
2. Identify the tasks that must follow one another (finish-to-start).
3. Identify the tasks that must follow one another partially but can begin around the same time (start-to-start).
4. Identify tasks that must be completed at the same time, before executing any subsequent tasks (finish-to-finish).

E

EARLY AND LATE START AND FINISH DATES

Every network diagram provides the basis for calculating the early and late start and finish dates for each task (Figure 12). The *early start date* is the earliest time that a task can begin, and the *early finish date* is the earliest time that a task can be completed. The *late start date* is the latest time that a task can begin, and the *late finish date* is the latest time a task can be completed. The early start and finish dates can be calculated by conducting the forward pass in a network diagram using the duration and lag of each task and the relationship type among tasks. The late start and finish dates can be calculated by conducting the backward pass in a network diagram using the duration and lag of each task and the relationship type among tasks.

GOALS

- Determine the start and stop dates for each task and an entire project.
- Calculate the total float and free float for each task.
- Determine the critical path.

OBSTACLES

- Failure to perform the forward and backward passes
- Failure to account for the lag and dependency relationships
- Failure to account for holidays and other nonworking days

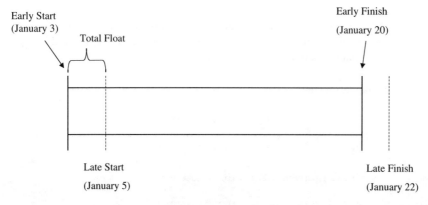

FIGURE 12 Early and late start and finish dates. (From Project Management Seminar presented by Practical Creative Solutions, Inc., 1996.)

STEPS

1. Identify all the relationships among tasks in a network diagram.
2. For each task, identify the duration, dependency type, and lag value.
3. Using the forward pass, calculate the early start and finish dates.
4. Using the backward pass, calculate the late start and finish dates.
5. Calculate the total float and free float.
6. Identify the critical path.

EARNED VALUE

Earned value is a set of equations to determine the overall cost and schedule performance of a project. The idea is to determine the difference between what was planned and what actually occurred, called *variance*. The equations are:

- *Actual cost of work performed* (ACWP): costs incurred during a specific period
- *Budget at completion* (BAC): total allocated budget
- *Budgeted cost for work performed* (BCWP): total budget for work completed plus portions of remaining tasks completed
- *Budgeted cost for work scheduled* (BCWS): total budget of tasks yet to be performed
- *Cost performance index* (CPI): determination of schedule performance efficiency
- *Cost variance* (CV): difference between planned and actual costs
- *Estimate at completion* (EAC): actual costs plus estimated costs to complete
- *Estimate to completion* (ETC): estimated costs of remaining work to be completed
- *Schedule performance index* (SPI): determination of schedule performance efficiency
- *Schedule variance* (SV): difference between planned and actual work
- *Variance at completion* (VAC): difference between total allocated budget and actual costs plus estimated costs to be completeed

The basic calculations are:

$$CV = BCWP - ACWP$$

$$SV = BCWP - BCWS$$

$$CPI\ (<1.0\ [negative]) = BCWP/ACWP$$

$$SPI\ (<1.0\ [negative]) = BCWP/BCWS$$

$$ETC = \text{cost of remaining work}$$

$$EAC = \text{cumulative actuals} + ETC$$

$$VAC = BAC - EAC$$

$$BAC = VAC + EAC$$

Goals

- Monitor the ongoing performance of a project from cost and schedule perspectives.
- Provide advance warning of any cost and schedule problems looming in the future.

Obstacles

- Lack of available data, expertise, and tools to calculate earned value
- Use of inaccurate or faulty data in calculations

Steps

1. Provide or obtain training on the concept of earned value.
2. Provide the tools, data, and other resources to calculate earned value.
3. Calculate earned value consistently and reliably.
4. Follow up on the results of the calculation.

E-MAIL

With the rise of Internet technology came a flood of e-mail. Managing a project is complicated enough without having to go through countless e-mail messages. Fortunately, the flood of e-mail can be successfully navigated. Actually, good e-mail etiquette is much the same as it is for memorandums. The content of all e-mail should be clear, concise, and accurate. It should contain identifiers (e.g., to, from, subject, date). It should have a purpose, such as a request for information or a call for action. In addition, it should list any courtesy copies to certain individuals. To ease the e-mail burden, consider developing templates for e-mails of a specific purpose. For records retention and ease of access, consider storing e-mail under various categories on the hard drives.

Goals

- Communicate information.
- Provide an audit trail.

Obstacles

- Unable to reduce the volume of e-mail
- Not setting up e-mail templates
- Writing unclear and wordy e-mail

STEPS

1. Develop some criteria for prioritizing e-mails.
2. Develop templates for different categories of e-mail (e.g., to inform or to take action).
3. Ensure that all e-mail contains the essential identifying information.
4. Recognize that some people hide behind e-mail to avoid more personal approaches for communicating.

EMOTIONAL INTELLIGENCE

Emotional intelligence, a concept developed by Daniel Goleman, is the ability to ascertain, understand, and apply one's emotions in a way that reflects maturity in dealing with circumstances on an intra- and interpersonal level. It requires becoming attuned to feelings and having the ability to deal with them in a mature, meaningful way which requires being trusting, open, and aware of emotions. A concept related to emotional intelligence is the idea of an adversity quotient (AQ), developed by Paul Stoltz. AQ indicates how well a person can deal emotionally with negative situations in relation to themselves and with others. The AQ determines a person's effectiveness and resiliency when responding to adversity.

GOALS

- Identify your emotional strengths and weaknesses and, if possible, those of other stakeholders.
- Capitalize on your own emotional strengths and compensate for your own weaknesses; if possible, do the same for others.

OBSTACLES

- Not applying all the principles behind the concept of emotional intelligence
- Assuming that people in general are emotionally intelligent

STEPS

1. Determine your level of emotional intelligence.
2. Observe the behavior of others and try to assess their emotional intelligence.
3. Use this knowledge to assign tasks and to determine teaming relationships that are consistent with the level of emotional intelligence in others.

ENNEAGRAM

The Enneagram is a personality analysis tool that identifies the essential nature of people and improves their awareness of the impact of their behavior on themselves and on others. The Enneagram is composed of a circle of nine points, which reveal the same number of personality types: perfectionist, giver, performer, tragic romantic, observer, devil's advocate, epicure, boss, and mediator. These personality types are revealed via three centers of intelligence: head, heart, and belly.

GOALS

- Identify the strengths and weaknesses of the personalities of stakeholders.
- Utilize people in a manner that capitalizes on their strengths and compensates for their weaknesses.
- Encourage solid teaming arrangements.

OBSTACLES

- Not applying all the principles behind the Enneagram
- Assuming that everyone can be categorized by one of the nine personality types

STEPS

1. Determine your personality type from the perspective of the three centers of intelligence.
2. Observe the behavior of others on your team from the perspective of the three centers of intelligence.
3. Use this knowledge to assign tasks and to determine teaming relationships that are conducive to the personality types of the individuals.

ENTITY-RELATIONSHIP DIAGRAMS

An entity-relationship (E-R) diagram is a graphical display of entities, their attributes, and their relationships with each other (Figure 13). An entity is an object,

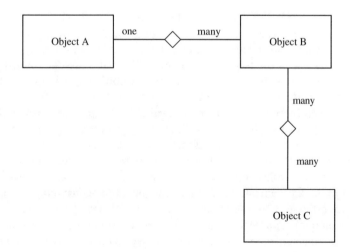

FIGURE 13 Entity-relationship diagrams.

or what is known as an *instance of a class*. Attributes are descriptive characteristics of an object, and their relationships reflect the cardinality between two objects (e.g., one to one, one to many). An E-R diagram enables identifying all the components of a system and discerning their relationships with each other. It serves as an excellent tool to use with a client to capture requirements and specifications as well as to lay the groundwork to develop a data model.

GOALS

- Draw a diagram showing the relationships among entities.
- Provide a basis for other subsequent models, such as a data model.
- Provide a means for capturing specifications and requirements.

OBSTACLES

- Not identifying all the objects and their relationships with one another
- Not using the diagram in subsequent models, such as for building a data model
- Not keeping the model current

STEPS

1. Determine the purpose and scope of an E-R diagram.
2. Identify its objects and the applicable relationships.
3. Identify the cardinality of the relationships.
4. Provide a legend and other identification information.
5. Place the drawing under configuration management to control changes.

ESTIMATING

Estimating is the process of determining how long each task will take to complete and indicates the time required for an entire project. Estimating proves useful in generating a meaningful schedule and calculating budget requirements. The work breakdown structure is the basis for estimating. Estimates are determined at their lowest level, known as the work package level. Upon completion of estimating, it is possible to perform summary-level calculations at various levels of the work breakdown structure. Several sources are available to tap to perform estimating, which is perhaps one of the most difficult jobs to perform on a large project because of the number of unknowns and the reluctance of people to commit themselves. Examples of such sources include historical records of previous projects, interviews with people having similar experiences, and access to databases of think-tank organizations. Perhaps the most time-consuming and grueling approach to estimating is the three-point, or PERT (Program Evaluation and Review Technique), approach, which involves determining three variables and results in an expected time to complete a project. Other estimating approaches exist but they tend to be mathematically overly complicated, assume that enough information is available, or skew results by allowing too much subjectivity.

Estimating

GOALS

- Give an idea of the time required to complete each task and an entire project.
- Determine in advance resource requirements for each task and an entire project.

OBSTACLES

- Failing to place parameters on estimates
- Striving for accurate estimates
- Failing to understand that estimates are a snapshot in time
- Not acknowledging the role of subjectivity in estimating
- Relying on estimates that are too optimistic or too pessimistic
- Not documenting reasons behind estimates
- Not using a consistent approach when estimating

STEPS

1. Determine the desired estimating approach (e.g., guess, PERT).
2. Obtain guidelines and inputs from stakeholders.
3. Perform calculations for time and cost.
4. Incorporate time estimates in the schedule.
5. Compare results with what is identified in the statement of work for the schedule and budget.
6. Review with stakeholders.
7. Revise estimates and recalculate.

F

FACILITATION

Facilitation is managing a working session or meeting to get results without influencing those results. A good facilitator focuses the energies of people toward a common goal and agenda while simultaneously maintaining a high tolerance for diversity. In other words, the facilitator must be a good steward. An effective facilitator, therefore, has good interpersonal, presentation, organizational, and problem-solving skills, as well as the ability to generate credibility, trust, and participation. Good preparation, too, prior to a session is the hallmark of an excellent facilitator.

GOALS

- Have the participation of all attendees.
- Have all sessions focus on goals and objectives by adhering to an agenda.
- Provide meaningful results, such as a solution to a problem.
- Have everyone leave with a win–win feeling.

OBSTACLES

- Failure to produce an agenda
- Sessions not focused on goals and objectives
- No advance preparation
- Failure to build rapport with attendees
- Inability to deal with conflict constructively
- Not applying good decision-making and problem-solving tools and techniques

STEPS

1. Before the session:
   a. Identify the most appropriate location.
   b. Prepare an agenda.
   c. Prepare for the facilitation in advance.
   d. Determine roles and responsibilities (e.g., facilitator, scribe).
2. During the session:
   a. Apply conflict-management skills.
   b. Apply good problem-solving skills.
   c. Focus on the goals and objectives.
   d. Take notes.

e. Give frequent breaks.
   f. Engender a sense of trust and credibility.
   g. Encourage ownership of results.
   h. Accept diversity of opinion.
   i. Avoid the position of becoming the decision maker.
   j. Apply good interpersonal skills.
   k. Apply good presentation skills.
   l. Apply effective and active listening skills.
3. After the session:
   a. Distribute minutes.
   b. Maintain dialog with participants.

FAST TRACKING

Fast tracking is accelerating the life cycle of a project by overlapping tasks (Figure 14). The goal is to deliver a product or service quickly. While fast tracking may provide many time-to-market advantages, it can be a risky approach filled with many dangers: loss of oversight when supervision of the work occurs at a high level; the temptation to make quick fixes that later in the life cycle can cause considerable problems and cost even more to fix; and, finally, burnout. To meet a tight schedule, people on a project team will likely work long hours because the workload is compressed into a shorter than usual duration. Burnout then leads to departures and mistakes.

GOALS

- Accelerate the project life cycle.
- Apply resources very efficiently.

OBSTACLES

- Overlooking key actions that would expedite the life cycle
- Increasing potential for rework

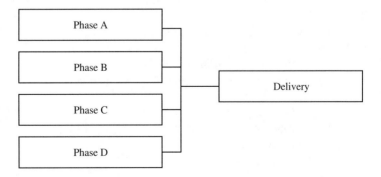

FIGURE 14 Fast tracking.

- Focusing on the quick fix
- Burnout

STEPS

1. Determine the reason for fast tracking.
2. Determine which phases to overlap.
3. Determine any potential risks that may occur and plan how to respond to them.
4. Recognize the potential negative impacts that could arise once the project concludes.
5. Prepare for those negative impacts.

FISHBONE DIAGRAM

The fishbone diagram is a graphical technique for determining the relationship between the causes of a problem and their results (Figure 15). The five groups of causes that contribute to a problem are material, manpower, machine, measurement, and method. In the fishbone diagram, each group reflects one or more contributors to a result. A central arrow is drawn at the top to reflect the overall effect of the causes, five adjacent arrows each represent a group, and more arrows represent contributors to each group.

GOALS

- Identify the main causes of a problem.
- Identify all the contributing factors.

OBSTACLES

- Not using all five groupings
- Not identifying all the contributors to a group

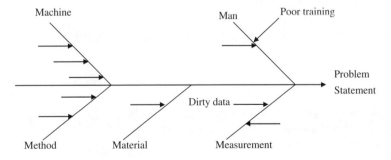

FIGURE 15 Fishbone diagram.

STEPS

1. Define the problem or issue to address.
2. List all the causes, and group them according to the categories of material, machine, measurement, man, or method.
3. Draw the fishbone.

(As an alternative, perform the first three steps with other people.)

4. Analyze the contents of the chart, noting major causes and their impacts (e.g., causing delays, rework, or duplicate effort).
5. Develop a plan for implementing the improvements.
6. Implement the improvements.

FORMS

A form is a medium used to capture data, store it, and deliver it to an appropriate destination. It can be in hard or electronic copy format. Forms can be used to capture a host of data on many project management topics, such as activity description, estimating, resource allocation, assignments, problem occurrence, change control, resource utilization, and purchasing. Ideally, a form should have the following characteristics: It should have a source and destination. It should contain simple instructions. It should be logically organized, have plenty of white space, contain only relevant requests for data, and require minimal effort to complete.

GOALS

- Make the form clear and concise.
- Capture only relevant, meaningful data in the form.
- Produce a form that is not time consuming to complete.

OBSTACLES

- Using different versions of the same form
- Capturing irrelevant data and information
- Incurring high maintenance costs
- Requiring too much data and information to be provided in a small area of a form

STEPS

1. Determine purpose of form.
2. Determine audience.
3. Determine source and destination.
4. Determine content to capture.
5. Determine disposition of form.
6. Determine storage medium.

FORWARD AND BACKWARD PASSES

Both passes are used to calculate the early and late start and finish dates for each task in a network diagram, a process that involves using the logical sequence of tasks, their durations and dependencies, and lag values to calculate the early start dates and then calculating the late start and finish dates using reverse logic.

GOALS

- Calculate early and late start and finish dates.
- Determine the critical path.

OBSTACLES

- Not understanding the purpose of the forward pass
- Not understanding the purpose of the backward pass
- Not applying the forward and backward passes to calculate dates

STEPS

1. Develop a network diagram.
2. Develop estimates for all tasks.
3. Move from left to right in the network diagram, calculating the early start and finish dates for each task.
4. Move from right to left in the network diagram, calculating the late start and finish dates for each task.
5. Note those tasks where the early and late start and finish dates match, thus indicating the critical path or paths.

FRAMEWORKS AND METHODOLOGIES

A framework and methodology provide a structured approach for developing deliverables and managing projects. The difference between the two is that a framework is at a higher level of abstraction than a methodology. Regardless, both should address topics such as the roles and responsibilities of stakeholders, processes and their respective inputs and outputs, and tasks necessary to produce desired results. All this information is integrated and documented. Frequently, both are used with an automated tool. Of course, no framework or methodology can guarantee success; they are only the tools to get results.

GOALS

- Provide a structure to manage a project.
- Manage more efficiently and effectively.
- Identify key deliverables.
- Provide a roadmap to navigate down the project life cycle.

OBSTACLES

- Adhering too rigidly or too lightly to a framework or methodology
- Applying an inappropriate framework or methodology based upon the context of the situation

STEPS

1. Identify requirements for a framework or methodology.
2. Perform a comparative analysis.
3. Select the most appropriate framework or methodology.
4. Train people on how to use it.
5. Allow time for learning curves.
6. Be patient.
7. Consider the context of the situation in which the framework or methodology will be used.

FUNCTIONAL HIERARCHY DIAGRAM

A functional hierarchy diagram is an overall approach to identifying the components of a system or product at different levels of abstraction and detail (Figure 16). Usually it is developed using a top-down approach, moving from the highest conceptual level down to the greatest acceptable levels of granularity; hence, it presents a parent–child relationship among different levels of modules, with the top level being the parent and the lower level being the child. Under some circumstances, a module can be both a parent and a child. Each module usually has a unique designator followed by a description about its function. A functional hierarchy chart lays the groundwork for subsequent modeling efforts (e.g., data flow diagramming).

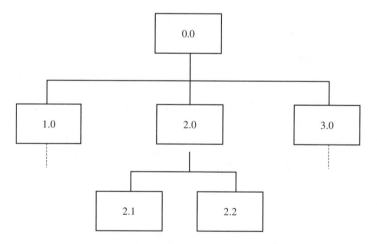

FIGURE 16 Functional hierarchy diagram.

Functional Hierarchy Diagram

GOALS

- Simplify the development of a product or service.
- Allow for reuse of certain components.
- Identify the importance of certain components.

OBSTACLES

- Exploding down to an excessive level of detail or a high meaningless level of detail
- Not using the chart as a basis for subsequent modeling exercises

STEPS

1. Take either a bottom-up or top-down approach (often the latter is taken).
2. Keep in mind how the contents of the diagram will be used.
3. Use some heuristic to determine when the diagram is complete.
4. Assign each object a unique identifier and description.

G

GLOBALIZATION OF PROJECTS

Thanks to the rise of international commerce, many projects in large companies use people from different countries. This diversity can provide great strength if project managers capitalize on it. It also presents some great challenges, as people on a multicultural team differ in beliefs, perspective, behavior, and work style. Dealing with verbal and nonverbal interaction as a project progresses through its life cycle can also be a challenge. For teaming purposes, however, it is imperative that project managers learn to take advantage of this diversity to achieve the vision of their project.

GOALS

- Capitalize on the strengths of a diverse workforce.
- Generate innovative ideas.

OBSTACLES

- Failure to appreciate the difficulty in dealing with cultural differences
- Failure to account for the impact of international events and legal complications

STEPS

1. Learn about the backgrounds of individual team members.
2. Recognize how to capitalize on the diversity to further project progress.
3. Acknowledge and deal with cultural differences.
4. Incorporate the impact of potential international events and legal situations during risk management.

GOAL

A goal is a statement of intent that clearly indicates the direction of a team. A goal is the basis for generating meaningful objectives. Quite often, a project has several goals supported by one or more objectives. The prime characteristic of a goal is that it is definable but not measurable. An example of a goal is "build a small state-of-the-art inventory management system within budget and by a certain date." A goal should be stated in one simple sentence and can be supported by measurable objectives.

GOALS

- Give the team a sense of direction.
- Effectively and efficiently execute a project.

OBSTACLES

- Not defining goals adequately
- Failing to generate commitment to goals
- Not recognizing the relationship between goals and objectives

STEPS

1. Identify the final result of the project.
2. Review any supporting documentation that might suggest goals.
3. List all the goals.
4. Group the goals by some criterion or criteria.

GOLDEN VS. IRON TRIANGLE OF PROJECT MANAGEMENT

The *iron triangle* of project management emphasizes the relationships among cost, schedule, and quality; the *golden triangle* of project management emphasizes the relationships among cost, schedule, quality, and people by placing people at the center of the iron triangle (Figure 17). People are the one element that ties the other elements together. Too often the emphasis is on the iron triangle to the exclusion of the people. The emphasis on people in the golden triangle helps maintain a balance among cost, schedule, and quality.

GOALS

- Encourage a balanced view for defining the criteria for success.
- Apply a more systemic view of managing a project.

OBSTACLES

- Failure to recognize the importance of people in meeting cost, schedule, and quality demands
- Inability to balance the components of cost, schedule, quality, and people
- Failure to integrate cost, schedule, quality, and people

STEPS

1. Define the goal or the project.
2. Determine the cost, schedule, quality, and people criteria necessary for success.
3. Develop all plans with these four components in mind.

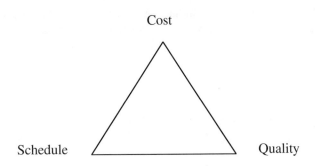

Iron Triangle of Project Management

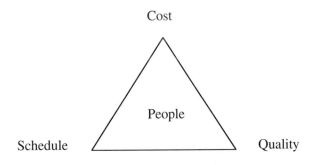

Golden Triangle of Project Management

FIGURE 17 Golden vs. iron triangle of project management.

4. During the execution of plans, constantly keep a balanced view of the components.
5. If one element becomes more important than the others, attempt to achieve a more balanced effect or compensate somehow for overemphasizing a component.

GROUPTHINK

Groupthink occurs when a group of individuals seek unanimity regarding the appraisal of particular circumstances. In other words, a group of people places extreme pressure on individuals to conform even to the point of eliminating free thought and expression to maintain harmony. The implications of groupthink are that people on a team filter information and skew their judgments for the sake of unanimity. The results are a failure to adequately assess reality and develop realistic solutions.

Goals

- Recognize the strong force of conformism associated with groupthink.
- Apply approaches that enable the free flow of ideas and communication.
- Implement approaches that prevent the occurrence of groupthink.

Obstacles

- Not recognizing groupthink when it exists
- Allowing groupthink to occur by not applying techniques to reduce its influence
- Failing to provide an environment that encourages the free exchange of ideas and information

Steps

1. Attempt to detect the presence of groupthink that might exist on a project.
2. Determine the degree of negative impact the groupthink might have.
3. Implement tools and techniques (e.g., brainstorming, emphasizing facts and data) to offset the effects of groupthink.

HERZBERG THEORY OF MOTIVATION

Frederick Herzberg developed what is known as the two-factor theory of motivation. The *maintenance factors* (or dissatisfiers) are very job centered and involve topics such as salary, job conditions, and policies. They dissatisfy when they are lacking in some form or another. The *motivational factors* (or satisfiers) do not dissatisfy if they are not present, but they can increase satisfaction when implemented. Motivational factors include concepts such as achievement and recognition.

GOALS

- Identify and apply the satisfiers on a project.
- Identify and mitigate the impact of dissatisfiers.

OBSTACLES

- Not distinguishing between satisfiers and dissatisfiers
- Emphasis on dissatisfiers over satisfiers and vice versa

STEPS

1. Review current policies, procedures, and practices for motivating team members.
2. Identify the maintenance and motivational factors having a positive affect and effect.
3. Identify the maintenance and motivational factors having a negative affect and effect.
4. For factors having a negative effect, determine what actions to take to lessen their impact or eliminate them.

HEURISTICS

Heuristics is an information processing approach that applies rules of thumb or guidelines to deal with situations. In other words, they work most of the time, but not 100%. They enable making quick decisions with minimal information.

Goals

- Develop a workable solution to a problem or overcoming an obstacle.
- Address a problem or obstacle quickly.

Obstacles

- Not being able to break away from trying to be precise
- Tending to strive for perfection
- Using heuristics when precision is necessary
- Not using heuristics when precision is unnecessary

Steps

1. Make a conscious effort to avoid trying to develop the perfect solution or find a precise answer.
2. Define the problem.
3. Develop several rules of thumb to solve a problem.
4. Select a rule of thumb that you think will best solve the problem.
5. Implement the rule of thumb.
6. Obtain feedback on the effectiveness of its implementation.

I

IMAGINEERING

Imagineering is an information processing approach to creation of an ideal form of an object, process, etc., capturing every detail. The idea is to visualize something perfect and use that image to develop a real-world equivalent.

GOALS

- Generate a perfect picture of what a deliverable might be.
- Experiment mentally with different ways to solve a problem or satisfy a requirement or specification.

OBSTACLES

- Assuming all details are equal
- Emphasizing being realistic over being idealistic
- Not allowing the imagination to take over
- Being influenced by other people's preferences
- Not being able to step away from one's own prejudices

STEPS

1. Clear your mind.
2. Find a comfortable, quiet place to sit.
3. Remove all preconceived notions.
4. Mentally construct perfect ideas, objects, or processes.
5. Decide on an idea, object, or process.
6. Focus on that idea, object, or process.
7. Capture on paper that idea, object, or process.
8. Implement that idea, object, or process.
9. Eventually compare what is built with the perfect image captured in your mind or on paper.

INFORMATION LIFE CYCLE

The information life cycle involves acquiring and processing data and creating information. It consists of seven phases: *identification phase*, determining what data are needed to generate information; *acquisition phase*, acquiring data to satisfy a goal and the needs of an audience; *organization phase*, obtaining the data and putting it in a meaningful order; *verification phase*, ensuring that the data are accurate; *interpretation phase*, reviewing the data and converting them to information; *presentation phase*, delivering information in a manner that satisfies the goal and needs of the audience; and, finally, *utilization phase*, using the information to achieve a goal.

GOALS

- Provide useful, accurate, and meaningful information.
- Use reliable data to convert into information.
- Provide an orderly way to deal with data overload.

OBSTACLES

- Failing to distinguish between data and information
- Using dated, inaccurate data to generate information
- Introducing bias when turning data into information
- Bypassing phases of the information life cycle

STEPS

1. Determine:
   a. Purpose of your information
   b. Audience
   c. Source of data
2. Obtain the data.
3. Check the data for reasonableness and validity.
4. Organize the data for manipulation.
5. Convert the data into information by applying various tools and techniques for data manipulation.
6. Place the information into a presentable form.
7. Ensure that the information satisfies the purpose and requirements of the audience.

INTERNAL RATE OF RETURN

The internal rate of return, also known as the marginal efficiency of investment, is used to determine what a new project will earn after all its costs have been paid. It measures the cost of capital borrowed from lending institutions. The key variables for calculating the internal rate of return are the expected return over operating expenses and the current interest rate. The formula is:

$$\frac{\text{First year rate over expenses}}{(1 + \text{internal rate of return})^1} + \frac{\text{N year return over expenses}}{(1 + \text{internal rate of return})^n}$$

$$\frac{\$17,000}{(1+0.8)^1} + \frac{\$20,000}{(1+0.8)^2} + \frac{\$23,000}{(1+0.8)^3} = \$55,554$$

GOALS

- Determine the rate of return of a new product or service.
- Determine whether to make an investment to build the product or deliver the service.

OBSTACLES

- Failing to recognize that the internal rate of return is only one method of analysis
- Using exaggerated or inaccurate data in variables

STEPS

1. Identify the following for each calculation:
   a. Return over operating expenses
   b. Anticipated rate
2. Calculate the internal rate of return for each scenario (e.g., project).
3. Chart the results of the calculations.
4. Select the desired option.

INTERVIEWING

Interviewing is meeting with one or more individuals to acquire data and information about a specific subject. The two basic types of interviews are *structured* and *unstructured*. Structured interviews require using a sequential set of questions to obtain a specific piece of data or information. Unstructured interviews use open-ended questions to give interviewees the latitude to respond in greater detail to a question.

GOALS

- Obtain data and information.
- Acquire a clear understanding of a subject.
- Solicit tacit and implied explanations about a subject.
- Use interviewing as a means for obtaining leads to gain additional data, information, or insights.

OBSTACLES

- Trying to influence the response of the interviewee
- Using the wrong type of interview approach for the desired results
- Picking the wrong time and place to conduct an interview

STEPS

1. Determine:
   a. Purpose of the interview
   b. Type of interview
   c. Interviewee
   d. Date, time, and place
   e. Method of recording notes
2. Conduct the interview.

INTUITION

Intuition is also known as "gut feel." It is an instinctive feeling that has little or no relevance to facts, data, or logic. Although rarely addressed in project management, it plays a significant role in the decision-making process because more often than not the facts and data are scarce or conflicting and little time exists to clarify or obtain more facts and data. Under such circumstances, a project manager must rely on a general sense that a decision made was the right one. As specific details become available, take corrective action, if necessary.

GOALS

- Take a calculated risk to seize an opportunity to achieve a goal or objective.
- Take decisive action rather than waiting for more information.

OBSTACLES

- Neglecting the importance of intuition in decision making
- Retreating into the collection of more facts and data rather than making intuitive judgments

STEPS

1. Listen to your instincts or pay attention to your gut feel.
2. Recognize that no matter how many facts and data are collected they will not ever be enough.
3. Understand that decision making has as much of an emotional or physical component as it does logic.
4. Seek the best solution, not the "right" solution.

ISO 9000

ISO 9000 is a set of international standards established to develop and implement a quality management system. The standards pinpoint specific activities that must be conducted to become ISO 9000 qualified (referred to as certification). ISO 9000 helps ensure that products and services do satisfy the intended customer requirements. Basically, ISO 9000 requires documenting processes (especially ones related to quality), maintaining adequate records of performance, and consistent implementation of processes.

GOALS

- Satisfy customer requirements.
- Document processes.
- Maintain records and data.
- Continue tracking ongoing compliance with ISO 9000.

OBSTACLES

- Holding incomplete knowledge about ISO 9000
- Misunderstanding the purpose and scope of ISO 9000
- Failing to identify the customers
- Misunderstanding the true requirements of customers

STEPS

1. Obtain copies of the applicable ISO 9000 documents.
2. Review or survey current practices in regard to quality.
3. Compare current practices to those described in the ISO 9000 documents.
4. Identify discrepancies.
5. Determine areas for improvement.
6. Prepare a plan for implementing improvements.
7. Prepare the necessary documentation.
8. Execute the plan, ensuring compliance with the documentation.
9. Consider conducting an ISO 9000 certification assessment by somebody not connected with the project.

ISSUE–ACTION DIAGRAM

An issue–action diagram is a decision-tree-like approach to determining specific courses of action for an issue (Figure 18). The idea here is that a specific issue requires selecting an action or alternative to resolve an issue.

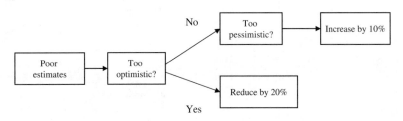

FIGURE 18 Issue–action diagram.

GOALS

- Reduce complexity in decision making.
- Provide a defined pathway for dealing with various scenarios.
- Reduce the extensive oversight of team members.

OBSTACLES

- Failing to identify all the relevant issues and actions
- Failing to identify all the relationships among issues and actions
- Having strings becoming so long as to become unmanageable
- Incorrectly assuming the relationships among issues and actions

STEPS

1. Determine:
   a. Purpose of the issue-action diagram
   b. Audience
2. Identify all issues and their corresponding actions.
3. From left to right (first issue, then actions) and from right to left (from last actions to preceding issues), draw an issue–action diagram.
4. Draw a box for each issue and action and a vector to show the relationships.
5. Ensure that all issues are connected to a given set or to more actions to take.
6. Ensure that no gaps exist in the diagram.

K

KEY CONTACT LISTING

A key contact listing is a list of individuals to notify during a project (Figure 19). These people may be contacted because of their responsibilities, expertise, or knowledge. Each entry includes the name, mail stop, phone number, address, company, e-mail address, and other relevant information. Often, it is part of the project manual on the project website. Because people are constantly coming and going on a project, it is imperative to keep the list current and accessible to everyone.

GOALS

- Maintain a listing of important people.
- Have ready access to people with required expertise or knowledge.

OBSTACLES

- Not keeping list updated
- Not being complete or accurate
- Not being accessible to stakeholders, especially project team members

Name	Expertise	Mail Stop	Phone Number	Company
Brahms, Robert	COBOL programming	X5-DC	(816) 455-0911	Endless Loop, Inc.

FIGURE 19 Key contact listing.

STEPS

1. Identify all the stakeholders.
2. Capture their expertise or roles.
3. Capture all the relevant contact information.
4. Keep the listing current by revisiting the contents periodically.
5. Publish the listing of all stakeholders.

L

LAG

Lag is the time of apparent inactivity that may occur between completion of a task and the start of tasks that immediately follow (Figure 20). Lag can be for any length or increment of time (e.g., hours or months). Lag can be assigned to all three dependency relationships: finish-to-start, start-to-start, and finish-to-finish.

GOALS

- Determine the "dead time" between tasks.
- Calculate the start and finish dates for each task and an entire project.
- Use this information to determine the critical path.

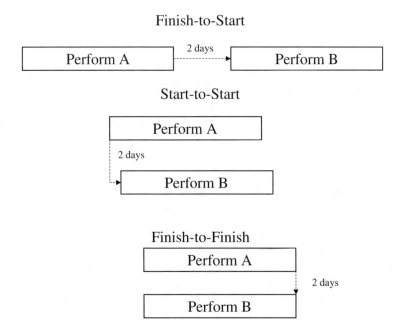

FIGURE 20 Lag. (From Project Management Seminar presented by Practical Creative Solutions, Inc., 1996.)

OBSTACLES

- Failure to identify lag time between tasks
- Failure to account for lag time in the calculation of start and stop dates for tasks

STEPS

1. Determine:
   a. Logical sequence of tasks
   b. Dependency type among tasks
   c. Amount of time of inactivity (lag value) in the relationships among tasks
2. Use the logical sequence, dependency type, and lag value to calculate the early and late start and finish dates for each task.

LATERAL AND VERTICAL THINKING

Edward de Bono developed the concept of lateral vs. vertical thinking. The premise behind lateral thinking is that people develop and use patterns when thinking. Vertical thinking, which is the traditional way of thinking, takes a linear, correct-answer approach to thinking; it is selective and analytical. Lateral thinking takes a different approach to thinking in that it requires recognizing people as self-organizing systems that reinforce patterned thinking. The idea is to break the patterns by emphasizing synthesis over analysis, being inclusive rather than exclusive, and looking for alternatives rather than seeking the correct answer.

GOALS

- Generate innovative ideas, tools, and techniques.
- Encourage creative thinking.
- Offset the negatives of specialized, routine thinking.

OBSTACLES

- Subscribing to an inflexible paradigm
- Always assuming that a question has only one correct answer
- Subscribing to a methodical, linear approach toward managing a project
- Not providing an environment in which creativity can flourish

STEPS

1. Identify patterned thinking, such as a prevailing paradigm, that might be followed on a project.
2. Identify some problems or obstacles on the project.
3. Determine different alternatives for solving each problem.
4. Experiment by playing around with each alternative.

5. Look at alternatives from various angles or perspectives.
6. Strive not to be methodical or mechanical in solving a problem.
7. Try to think cross-functionally.
8. Step outside the boundaries.
9. Remove preconceived notions.
10. View situations as matters of gray rather than black and white.

LEADERSHIP SKILLS

Much has been written about the subject of leadership in general and leadership qualities in particular. Two of the biggest names in the study of leadership are Warren Bennis and Barry Posner, who have provided considerable insight on the subject. Typically, a project manager, as a leader, is interpersonal (e.g., applies active listening), communicative, decisive, objective, committed, motivational, honest, consistent, visionary, and proactive. A project manager should exhibit these four general qualities: being interpersonal, modeling, being communicative, and exhibiting team building.

GOALS

- Assemble a team focused on the particular vision, goals, and objectives.
- Engender high morale and *esprit de corps* among stakeholders.

OBSTACLES

- Failing to recognize the leadership role of the project manager
- Not distinguishing between project management and project leadership

STEPS

1. Assess the extent to which the four criteria provided above are being applied by stakeholders in a leadership position in a project.
2. Take corrective action, if necessary.

LEADERSHIP: COMMUNICATING SKILLS

Communicating involves informing all stakeholders about the vision and tasks of a project and encouraging feedback. Some of the ways to achieve such communication on a project is through reports, forms, newsletters, meetings, and procedures.

GOALS

- Encourage teaming.
- Reduce miscommunications and misunderstandings.
- Reduce negative conflict.

OBSTACLES

- Emphasizing speaking over listening
- Emphasizing electronic ("distant") communications over interpersonal communications
- Building rapport through empathy, not sympathy

STEPS

1. Listen *and* hear.
2. Communicate constantly and consistently.
3. Keep an open mind.
4. Share information.
5. Welcome both good and bad information.
6. Do not shoot the messenger.
7. Empathize rather than sympathize.

LEADERSHIP: INTERPERSONAL SKILLS

Interpersonal skills include being able to communicate and empathize with many people. These skills are essential for motivating team members and other stakeholders. Good interpersonal skills, at a minimum, require the ability to be sensitive to the feelings of others and to solicit and be receptive to feedback.

GOALS

- Encourage greater communications.
- Improve teaming.
- Reduce opportunity for negative conflict.

OBSTACLES

- Lacking empathy for stakeholders
- Hearing but not listening
- Emphasizing managerial or technical aspects of a project over the people involved

STEPS

1. Recognize that formal and informal relationships play important roles on a project.
2. Listen as well as hear.
3. Empathize, not sympathize.
4. Follow up on promises, commitments, etc.
5. Build and maintain trust and credibility.

LEADERSHIP: MODELING SKILLS

Modeling is exhibiting desirable characteristics when managing projects. It is commonly referred to as "walking the talk." The idea is that a person is considered reliable and trustworthy because he merges expectations of behavior with reality. Consistency in behavior is also essential for building credibility.

GOALS

- Build trust through consistent behavior.
- Set the stage for future behavior.

OBSTACLES

- Not fulfilling the role expectations set by stakeholders
- Emphasizing project management over project leadership

STEPS

1. "Walk the talk."
2. Maintain consistency when managing a project through constancy of behavior.

LEADERSHIP: TEAM BONDING SKILLS

Team bonding involves engendering and maintaining a high *esprit de corps* throughout the life cycle of a project. For team bonding to occur, stakeholders must focus on a vision and collaborate with one another to achieve it. Formal and informal techniques (e.g., achievement awards and picnics, respectively) can be used to encourage and maintain team bonding.

GOALS

- Ensure greater collaboration and cooperation.
- Build *esprit de corps*.
- Encourage greater sharing of information.
- Reduce negative conflict.

OBSTACLES

- Emphasis technical and managerial aspects of a project over people considerations
- Lack of focus on the goals and objectives
- Lack of commitment to deliverables
- Poor resolution of conflict

STEPS

1. Maintain ongoing communications.
2. Reward the team as well as individuals.
3. Focus all team activities and efforts on goals and objectives.
4. Locate team members as close as possible.
5. Encourage sharing of information and feelings.

LEADING

Leading is one of the four major functions of project management and the only one that occurs simultaneously with others throughout a project life cycle. Leading motivates people to perform satisfactorily in a way that inspires others to accomplish the vision, goals, and objectives at a level that exceeds expectations. It consists of three elements: project leadership, leadership qualities, and team building. *Project leadership* involves performing six responsibilities: providing vision, communicating, maintaining direction, motivating, being supportive, and team building. *Leadership qualities* involve having expertise in these skill areas: interpersonal relations, modeling (e.g., behavior), communications, and team bonding. *Team building* involves instituting effective span of control, assigning roles and responsibilities, building team spirit, providing effective communications, applying the unity-of-command principle, and engendering a supportive environment.

GOALS

- Accomplish vision, goals, and objectives efficiently and effectively.
- Have a highly motivated project team.

OBSTACLES

- Not understanding the difference between leading and managing
- Failing to understand the leadership role of a project
- Not recognizing the leadership role of other stakeholders

STEPS

1. Provide leadership.
2. Develop leadership qualities.
3. Encourage team building.

LEADING: BEING SUPPORTIVE

Being supportive is providing value-added support to team members when achieving a vision. It involves providing the right tools; engendering a good, positive work environment; and overcoming obstacles to productivity.

Goals

- Allow a team to focus on getting the job done.
- Isolate a team from extraneous influences.

Obstacles

- Failure to isolate the team from company politics
- Failure to provide providing adequate resources
- Failure to resolve differences among team members

Steps

1. Provide the team with the necessary resources to perform tasks.
2. Protect the team from political and other extraneous influences, if possible.
3. Facilitate collaboration.

LEADING: COMMUNICATING

Communicating is informing all stakeholders about the vision and tasks of a project and encouraging feedback. Some of the ways to achieve effective communication on a project is through reports, forms, newsletters, meetings, and procedures.

Goals

- Encourage teaming.
- Reduce miscommunications and misunderstandings.
- Reduce negative conflict.

Obstacles

- Emphasizing speaking over listening
- Emphasizing electronic ("distant") communications over interpersonal communications
- Failing to build rapport through empathy, not sympathy

Steps

1. Listen *and* hear.
2. Communicate constantly and consistently.
3. Keep an open mind.
4. Share information.
5. Welcome both good and bad information.
6. Do not shoot the messenger.
7. Empathize rather than sympathize.

LEADING: MAINTAINING DIRECTION

Maintaining direction is remaining consistent when executing the vision of a project. Some of the relevant project management disciplines that maintain direction are tracking, monitoring, and status collection and assessment.

Goals

- Provide a sense of confidence.
- Reduce opportunity for scope creep.
- Provide opportunities for greater collaboration.
- Increase efficiency and effectiveness.

Obstacles

- Not focusing on the vision
- Not communicating the vision
- Not obtaining regular feedback on performance

Steps

1. Maintain focus on the vision, goals, and objectives.
2. Maintain pulse on progress through status collection and assessment.
3. Take corrective action, if necessary.

LEADING: MAKING EFFECTIVE DECISIONS

Making effective decisions involves using sound managerial tools and techniques to determine a specific course of action. It requires converting data into information, using specific criteria to determine and weigh alternatives, selecting a course of action, taking action, and collecting feedback to take corrective action.

Goals

- Encourage efficient and effective results.
- Provide a sense of direction.
- Build confidence.

Obstacles

- Being indecisive or too quick to make a decision
- Not considering the most important elements of information
- Not obtaining commitment to a decision

Steps

1. Adopt a decision-making approach (e.g., PDCA [plan, do, check, act] cycle).
2. Ensure constant and reliable flow of data and information.

3. Recognize and avoid biases during decision making.
4. Maintain focus on the vision, goals, and objectives.

LEADING: MOTIVATING

Motivating is encouraging people to participate actively to attain the vision of a project. It involves making decisions, providing incentives, structuring jobs, encouraging team participation, and delegating.

GOALS

- Cultivate *esprit de corps*.
- Augment morale.
- Build commitment and ownership.

OBSTACLES

- Concentrating too much on the material aspects of motivation
- Relying too much on negative incentives
- Not using the formal and informal powers available to a project manager

STEPS

1. Determine availability of positive and negative incentives.
2. Meet with individual team members to determine needs, wants, etc.
3. Identify ways to reward for outstanding performance on both individual and team levels.

LEADING: PROVIDING VISION

Vision, from a project management perspective, is providing well-defined goals and objectives and a path to achieve them. If good project management disciplines are applied and followed, a vision will be revealed in a statement of work, work breakdown structure, schedule, and budget. Throughout the life cycle of a project, all information and activity should be evaluated from the perspective of achieving a vision.

GOALS

- Encourage a sense of confidence.
- Provide a sense of direction.
- Build *esprit de corps*.

OBSTACLES

- Lack of commitment to the vision
- Failure to communicate a vision

- Lack of focus on a vision
- Lack of clarity in a vision

STEPS

1. Develop a vision.
2. Communicate a vision.
3. Generate commitment to the vision.
4. Abide by a vision by incorporating it in the plans and actions of the project.
5. Revisit a vision.

LEADING: USING DELEGATION PROPERLY

Using delegation properly is a willingness to share workload without foregoing responsibility. It requires conducting a good assessment of people, knowing when to become involved and when to pull back, and providing effective feedback.

GOALS

- Encourage a sense of ownership and commitment.
- Build trust.
- Prioritize workloads.

OBSTACLES

- Dictating how to perform tasks
- Not following up on delegated tasks
- Not recognizing that a project manager can delegate work but not responsibility

STEPS

1. Determine what tasks to delegate.
2. Determine the skills, experiences, etc. necessary to perform delegated tasks.
3. Select people to perform tasks.
4. Explain the expectations for completion.
5. Perform follow-up on the progress of tasks, noting positive results and opportunities for improvement.

LEARNING CURVE

A learning curve is essentially the time required for a person to become effective when applying a new skill or knowledge to a task. People vary in learning speeds and capacities, which makes it very difficult to determine the impact of individual learning curves. Factors affecting the shape of the learning curve include the quality of the work environment, the complexity of skill or knowledge, previous related experience, talents, and availability of tools. Learning curves should be accounted for when making time estimates as they can impact costs and schedules. Learning curves also impact quality. It is imperative, therefore, to account for them when estimating time, costs, and schedules.

GOALS

- Reduce learning curve.
- Improve speed of applying new skills and expertise.
- Reduce rework due to errors and lack of knowledge or expertise.

OBSTACLES

- Failing to appreciate the impact of learning curves on individual performance
- Not incorporating learning curve considerations in time estimates
- Losing patience with people having long learning curves

STEPS

1. Incorporate learning curve considerations in time estimates.
2. Have patience with people applying new knowledge or skills.
3. Recognize that learning curves improve over time.
4. Look for opportunities to offset the negative impacts of learning curves.

LEARNING STYLE

People learn differently; however, learning generally goes through four stages, according to the Learning Style Indicator (McBer and Company). During *concrete experience*, the emphasis is on feelings, social interaction, and experience. *Reflective observation* recognizes that people learn different ideas from multiple perspectives and observation. *Abstract conceptualization* is thinking that uses systematic analysis and intellectual understanding. *Active experimentation*, the fourth stage, emphasizes action by influencing people and tackling a problem. People generally have a preferred stage, and progress through each one varies with each individual. In summary, learning follows this progression: feeling, watching, thinking, and doing.

GOALS

- Reduce learning curves.
- Encourage involvement in learning a new idea, tool, or technique.

OBSTACLES

- Bypassing the stages of learning
- Assuming that people learn at the same pace or in the same way

Steps

1. Recognize that everyone learns differently in terms of style and pace.
2. Remember that learning often involves a holistic or total experience.
3. Appreciate that learning improves with time.
4. Recognize that the learning curve plays an important role in affecting the application of skills on a project.
5. Encourage people to become emotionally and mentally involved when learning.

LEFT AND RIGHT BRAIN THINKING

According to psychological theory, the brain is partitioned into two halves, frequently referred to as the left and right hemispheres. While this concept is an oversimplification, it does provide some insight into why certain perspectives dominate on most projects. Hemispheric thinking is also known as brain dominance, and it manifests itself in the way projects are managed. Project managers who are left-brain dominant tend to emphasize logic, sequence, linearity, concrete thinking, scheduling, organization, written or verbal instructions, analysis, and following a known path. Project managers who are right-brain dominant tend to emphasize intuition or hunches, emotions, nonlinearity, abstract thinking, exploration, impulsive action, patterns, synthesis, visual presentations of material, and risk taking.

Goals

- Provide a balanced approach toward managing a project.
- Maintain a focus on the overall picture of a project without overlooking important details.
- Respond effectively during the life cycle of a project.

Obstacles

- Overemphasizing one brain dominance over another
- Failing to recognize the importance of seeing the overall picture without overlooking important details
- Not recognizing when brain dominance might be useful

Steps

1. For any project, evaluate the overall approach to managing it by identifying the concepts, knowledge, tools, and techniques that have been applied.
2. If the effects of a particular brain dominance seem to be extreme, consider balancing the dominance by applying concepts, knowledge, tools, and techniques associated with the other hemisphere.
3. Recognize that certain stakeholders may also have skewed brain dominance.
4. Recognize that brain dominance is not necessarily bad or good.

LESSONS LEARNED

A lesson learned is a document containing a postevaluation of a project that cites successes, problems, and future opportunities. Some typical sources of information for this document are interviews with stakeholders, statistical compilations, minutes, schedules, cost records, change control requests, and a statement of work and its subsequent revisions.

Goals

- Produce a document that serves as a historical biography of a project.
- Allow future projects of a similar nature to benefit from the experience.

Obstacles

- Failing to be thorough
- Allowing the document to become a political essay
- Pointing fingers
- Allowing the document to become unavailable

Steps

1. Prepare the lessons learned document toward the end of a project or immediately after the conclusion.
2. After the first draft, circulate the document for review.
3. Revise the draft and produce a finished document.
4. Store the document in a manner that makes it accessible to future project managers.

LEVELING

Leveling is removing extreme peaks and valleys in a histogram to optimize the use of project resources (Figures 21 and 22). Frequently, it is applied to labor resources to produce a realistic portrayal of their deployment.

Goals

- Provide for efficient and effective use of resources.
- Reduce negative impacts on team performance.
- Concentrate on production rather than concerns about hiring and firing people.
- Maximize the productivity of resources.
- Function proactively, not reactively.

Obstacles

- Not generating a meaningful histogram
- Failing to use the critical path, task relationships, and availability of resources when leveling

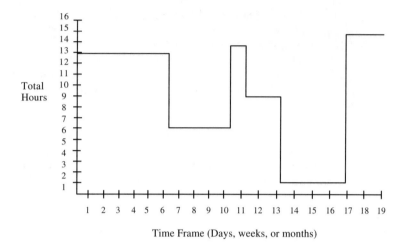

FIGURE 21 Unleveled histogram.

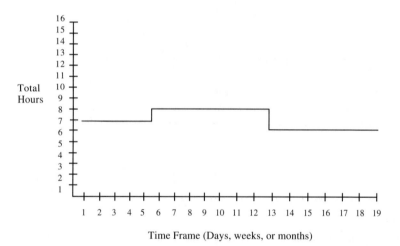

FIGURE 22 Leveled histogram.

- Not accounting for various ways to level a histogram (e.g., adding more people to a task or extending the duration of a task by allotting fewer hours per day to work on it)
- Accepting extensive use of overtime over a significant period of time despite flat peaks

STEPS

1. Explore these various options for leveling a histogram:
   a. Change dependencies among tasks.
   b. Lengthen the lag-time value between two or more tasks.

c. Extend the duration (flow time) of a noncritical task while reducing proportionately the hours per day to work on it.
d. Eliminate tasks from the schedule.
e. Change logic relationships among tasks.
f. Reduce the hours to perform a task and compensate for the reduction by purchasing equipment or pursuing some other alternative.
g. Reduce the number of hours per day to work on a task and its cumulative total number of hours.
2. Select the best option that reduces overtime and does not create a bottleneck in productivity.

LISTENING AND HEARING

Hearing is paying attention to the sound of a speaker's voice. Listening is hearing the sound but also picking up both the verbal and nonverbal meaning of the message. Listening is a skill often applauded but seldom exercised. Good listeners listen to what is and is not said. They pay attention to the words of the speaker and such nonverbal language as facial expressions. Good listeners know that they send verbal and nonverbal messages back to the speaker. They exhibit good eye contact, send physical and verbal cues to convey acknowledgment, ask meaningful questions, and are attentive. Above all, they recognize that listening is really two-way communication.

GOALS

- Obtain meaningful information about a topic or issue.
- Acquire an objective perspective.
- Empathize, not sympathize, with the speaker.

OBSTACLES

- Using selective listening
- Not distinguishing between what to hear vs. need to hear
- Failing to listen to the total message, which includes both words and body language
- Not recognizing the impact of one's own verbal and physical communications on the speaker

STEPS

1. Listen to what is and is not said and how it is said.
2. Get to the real meaning of the message.
3. Be attentive and reflective.
4. Empathize rather than sympathize with the speaker.
5. Be mindful of your own verbal and nonverbal messages and cues.
6. Use mannerisms and speech to acknowledge that you are listening.

LOGICAL AND PHYSICAL DESIGNS

Essentially, the two basic models are logical and physical. A logical data model is a graphic or narrative description of how a design will appear from a conceptual, functional perspective. A physical design is a graphic or narrative description of how a design will be implemented. Whether logical or physical, a good design is one that is very modular (that is, is broken into discrete components) and has well-identified interfaces. What makes each model unique is the organization of the components. Ideally, the logical model should be completed first, and then the physical model follows. The logical model serves as the ideal solution and the physical one represents the real-world manifestation of the former. Often, several logical and physical models are produced.

Goals

- Reduce the tendency to jump to the physical solution before determining the logical alternatives.
- Develop the most efficient and effective design.

Obstacles

- Jumping to the physical design before developing a logical equivalent
- Not tying the logical design to the physical design
- Failing to update the logical design when the physical design has changed

Steps

1. Develop the logical models first.
2. Develop the physical models last.
3. Pick the most appropriate physical model that best satisfies the chosen logical model.
4. Remember to change the logical model whenever the physical model changes and vice versa.

M

MANAGERIAL GRID

Robert Blake and Jane Mouton developed the theory that leadership style is a combination of a concern for people and production (Figure 23). The combination of these two variables creates five basic leadership styles: *impoverished management*, when the manager exerts very little effort over people or production; *task management*, when the manager emphasizes production over people (e.g., efficiency); *country club management*, when the manager emphasizes people over production (e.g., being supportive and considerate); *middle of the road management*, when the manager balances the need for production with the concern for people; and *team management*, when the manager integrates people and production, which is of primary importance in the project environment. Each type of leadership style has implications on what approaches will be taken to achieve results, the priorities being set, how conflict is resolved, the way motivational factors are applied, and how change is handled.

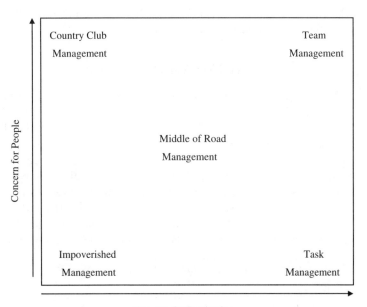

FIGURE 23 Managerial grid.

Goals

- Apply the appropriate leadership style.
- Identify ways to satisfy production and people concerns.
- Determine the positive and negative impacts with each leadership style.

Obstacles

- Emphasizing one leadership style when another would be more appropriate
- Failing to recognize the positive and negative aspects of the leadership style being applied
- Lacking a willingness to try something different

Steps

1. Assess the environment of the project and determine whether the emphasis is on people or production, or some balance of both.
2. Conduct the assessment by using criteria relating to practices, priority setting, conflict resolution, managing people, etc.
3. If the emphasis is on production, determine the negative impacts and implement solutions to lessen the negative influence.
4. If the emphasis is on people, determine the negative impacts and implement solutions to lessen the negative influence.

MASLOW HIERARCHY OF NEEDS

The needs hierarchy, developed by Abraham Maslow, bases human motivation on the premise that the behavior of humans is based upon satisfying one's needs (Figure 24). These needs, in turn, have a hierarchical arrangement. The five levels in sequence are physiological, safety, social, esteem, and self-actualization. Physiological needs are concerned with satisfying biological functions, such as hunger, sex, etc. Safety needs pertain to physical and psychological safety. Social needs entail one's involvement with other people, from affection to acceptance. Esteem needs are related to one's feeling of self-worth or importance, either with oneself or through others. Self-actualization needs are met when a person performs according to natural capabilities, talents, etc. People who reach this stage often describe satiation of this need as being a peak experience. Denial of any of these needs will tend to cause frustration, thereby increasing tension.

Goals

- Provide appropriate inducements to perform.
- Direct behavior toward positive pursuits for the benefit of the people involved as well as the entire project.
- Tap into the power of striving to satisfy self-actualization needs.

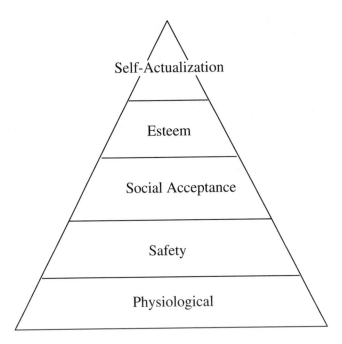

FIGURE 24 Hierarchy of needs.

OBSTACLES

- Misdiagnosing the real needs of a stakeholder
- Applying the idea of hierarchical needs as a law of human behavior

STEPS

1. Recognize that each person on a project may be at a different level in the hierarchy.
2. Try to ascertain the level of need for each individual on the team.
3. Identify ways to satisfy those needs, if possible.
4. Monitor the results and take corrective action, if necessary.

MATRIX AND TASK FORCE STRUCTURES

Two basic types of team structures exist: matrix and task force (Figures 25 and 26). In the *matrix structure*, people support multiple projects. It provides flexibility for responding to different market conditions and sharing people who have rare, specialized skills. A matrix structure is often used in industries with tight labor conditions and dynamic market conditions. In the *task force structure*, people support one project. This arrangement enables people to dedicate their time, attention, and expertise to achieve the goals and objectives of a particular project. Upon project completion, the team disbands. A task force structure is often used in industries with stable market conditions and sufficient supply of people with the requisite skills.

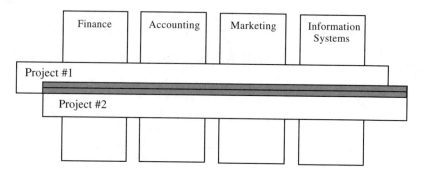

FIGURE 25 Matrix structure. (From Project Management Seminar presented by Practical Creative Solutions, Inc., 1996.)

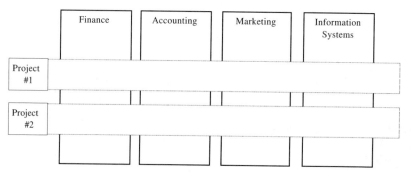

FIGURE 26 Task force structure. (From Project Management Seminar presented by Practical Creative Solutions, Inc., 1996.)

GOALS

- Apply people resources efficiently and effectively.
- Provide flexibility when responding to different environments.

OBSTACLES

- Common problems for matrix structure:
  - Conflicting priorities
  - Complex vertical and horizontal relationships
  - Information confusion
- Common problems for task force structure:
  - Temporary existence
  - Lack of available expertise

STEPS

1. Ascertain environment of the project.
2. Ascertain how other projects of a similar nature have been structured in the past.

Matrix

3. Adopt the appropriate structure.
4. Determine measures for countering or adapting to problems associated with each structure.

MATRIX

A matrix is a compact approach to displaying the relationships among various types of data. The intersection between two types of data is a cell that represents the value of a relationship. Sometimes a matrix can be three-dimensional, with a cell representing the relationship among all three data.

GOAL

- Efficiently and compactly display relationships among data.

OBSTACLES

- Failing to identify all relationships among types of data
- Creating a matrix that contains too much data, thereby becoming to unwieldy
- Failing to clearly mark or provide a legend for the matrix

STEPS

1. Determine:
   a. Purpose of the matrix
   b. Audience
2. Draw the matrix; include a y-axis and an x-axis, as well as gridlines to create cells to reflect the relationships between elements located on the two axes.
3. Determine the symbol to be used to reflect the value of any given relationship between elements.
4. Complete the matrix with titles, subtitles, and a legend.

MEAN, MEDIAN, AND MODE

The mean, median, and mode are means to determine the characteristics of a distribution of data. The *mean* takes the average of the observations of a group of data with a normal distribution. It uses this formula:

$$(\Sigma x)/n$$

where Σx is the sum of all the values of the data (x), and n is the total number of values. For example, if $\Sigma x = 50$, then $50/10 = 5$, which is the mean.

The *median* takes the middle value of a distribution of data. It requires sorting the data values according to magnitude and then picking the middle value of the rank order. For example,

5_10_1_20_25_30_35

The median value, then, would be 20.

Mode is used for very asymmetrical distributions. It requires choosing the value that occurs most frequently. For example,

2_3_4_3_3_

The mode in this case would be 3.

GOALS

- Draw meaningful conclusions from data.
- Assess the subsequent impact of any changes.

OBSTACLES

- Having insufficient data to draw any meaningful conclusion from the calculations
- Applying the wrong calculation (e.g., use mean rather than mode)

STEPS

1. Collect the data or use what is available.
2. Sort the data.
3. Determine the appropriate calculation based upon the amount and characteristics of data collected.
4. Draw a conclusion from the calculations.
5. Over time, display the results of numerous cycles of calculations in a graph or chart to reveal trends.

MEETINGS

Meetings are group sessions that result in efficiently and effectively accomplishing the goals and objectives of a project. For a typical project, three types of meetings are held: checkpoint review, status review, and staff. *Checkpoint review* meetings are held at key milestone dates in a project to assess the progress made up to a point in time and determine whether or not to proceed. *Status review* meetings are held regularly to review progress against schedule, cost, and quality criteria. *Staff meetings* are held regularly to discuss general issues and share information.

GOAL

- Conduct efficient and effective meetings.

OBSTACLES

- Not having an agenda
- Not adhering to an agenda

- Allowing meetings to run too long
- Allowing certain individuals to dominate
- Restricting discussion
- Not taking minutes

STEPS

1. Determine purpose of meeting.
2. Determine type of meeting (e.g., checkpoint review).
3. Determine attendees.
4. Determine location.
5. Determine supplies.
6. Prepare and distribute an agenda.
7. Conduct meeting by sticking to agenda.
8. Prepare minutes or notes.

MEMO

A memo, or more formally a memorandum, is a written instrument for capturing an idea, opinion, or result. A well-written memo is timely, clear, and concise. Any stakeholder can prepare one, and it should be easily accessible by anyone needing it. A memo can be prepared after a meeting, to raise an issue and force its resolution, to clarify a policy, to communicate important information, to schedule events, or to document an occurrence of an event.

GOALS

- Provide a record of results.
- Instill commitment.
- Provide traceability.
- Address and resolve issues.
- Provide a vehicle for communicating.

OBSTACLES

- Wording not clear, concise, or logical
- Lack of pertinent information
- Lack of formatting (e.g., date, addressee, signature block)
- Not easily accessible

STEPS

1. Determine the subject.
2. Determine the readers.
3. Prepare formatting (e.g., date, addressee, signature block).
4. Draft contents by initially preparing an outline, if necessary.

5. Edit the memo, looking for content, clarity, timeliness, and logic.
6. Revise the memo.
7. Publish the memo.

MEMORIZATION

The amount of information and documentation involved in most projects is staggering. Few people can effectively and efficiently deal with the volume of both. However, it is important that project managers and other stakeholders have the capacity to remember important information without having to refer back to reams of documentation that may not be readily available. Good memorization techniques can help overcome this challenge. These memorization techniques include recognizing and developing patterns; using analogies, metaphors, associations, comparisons, and contrasts; paraphrasing; and rhyming, among others. It is important to recognize that memorization involves more than filling one's brain with information. It requires repetitively reviewing the information and applying it to offset the inevitable decline in the ability to recall the information.

GOALS

- Recall important details about a topic or issue.
- Reduce effort and flow time to address a topic or issue.
- Avoid making costly mistakes and, consequently, extensive rework.

OBSTACLES

- Trying to memorize all information about a topic or issue
- Incorrectly memorizing the information
- Failing to distinguish important from unimportant information

STEPS

1. Identify what needs to be memorized.
2. Use various tools and techniques for memorization.
3. Recognize the importance of revisiting information and ideas occasionally to reinforce memory.
4. Periodically verify what has been memorized and apply it.

MENTORING VS. COACHING

Coaching is an important task of a project manager. It provides immediate feedback on performance and gives advice on developing a person's skills. It requires building a trusting, meaningful relationship with team members to help them progress. To do that, a project manager must generate credibility and respect, understand the needs and aspirations of team members, listen patiently, and be supportive. Mentoring has a much more long-term perspective than coaching and, therefore, is not associated much with the role of project manager. Mentoring involves giving advice and counseling on career choices beyond the life of a project. The goals, obstacles, and steps described below pertain to coaching only.

Goals

- Encourage effective and efficient performance by team members.
- Generate a sense of ownership and commitment by team members.
- Increase morale.
- Encourage a partnership between the project manager and team members.

Obstacles

- Attempting to "tell" rather than "sell"
- Becoming critical of people
- Failing to build trust or credibility with team members

Steps

1. Recognize the difference between coaching and mentoring.
2. For coaching, try to:
   a. Build a sense of partnership.
   b. Avoid judgmental comments.
   c. Remain objective.
   d. Rely on facts and data, not assumptions.
   e. Recognize the aspirations of the individual.
   f. Recognize your own and others' strengths and limitations.
   g. Build a sense of trust, integrity, and comfort.
   h. Keep in mind the culture and climate and their impact on the ability to coach.

METRICS

Metrics are measurements focused on the performance of a process or project. Metrics involve defining what to measure, collecting data, conducting measurements, interpreting results, and taking corrective action, if necessary. It is important to ensure that metrics serve a purpose and are not being collected for the sake of simply generating metrics. The best way to ensure that they serve a purpose is to identify key performance indicators or critical success factors. After establishing the metrics, it is important to collect, analyze, and act upon the data consistently and reliably. Metrics can be developed for cost, schedule, quality, and people performance.

Goals

- Identify opportunities for continuous improvement.
- Maintain focus on critical success factors.
- Reduce defects and rework.

OBSTACLES

- Failing to perform data collection, measurement, and interpretation activities on a consistent and reliable basis
- Introducing bias in calculated results
- Having metrics that provide little or no value

STEPS

1. Identify the key process areas, such as cost, schedule, or quality.
2. Determine the type of metric.
3. Determine what data to collect.
4. Collect data.
5. Compile the data.
6. Calculate results.
7. Identify any anomalies.
8. Present the information in the applicable format (e.g., tabular report or graphic).

MIND MAPPING

Mind mapping is a conceptual diagramming technique developed by Tony Buzan to take advantage of the mental capabilities of the brain to generate new ideas (Figure 27). The real power behind mind mapping is to employ the associative capabilities of the brain as they relate to a specific idea. The results are a diagram that clearly defines an idea and identifies the prioritized linkage of concepts behind it.

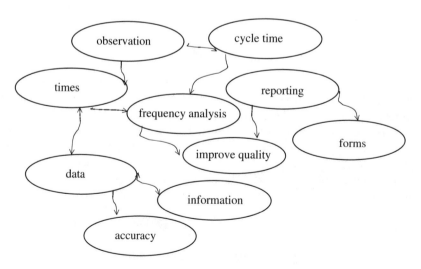

FIGURE 27 Mind mapping.

Modeling

GOALS

- Identify concepts, ideas, etc.
- Map the relationships among these concepts, ideas, etc.

OBSTACLES

- Thinking in terms of flow of control
- Inability to free one's mind
- Making incorrect associations among elements of an idea or concept

STEPS

1. Clear your mind.
2. Find a quiet, comfortable place to sit.
3. Discard all preconceived notions.
4. Determine the subject of study.
5. Using a large sheet of paper, write the main idea in the center.
6. Record any ideas, characteristics, etc. that come to mind and draw circles around each one (although this is not necessary).
7. Draw lines to reflect relationships among ideas, characteristics, etc.
8. Avoid the tendency to fall into an approach of flow of control (e.g., logical sequence) when mind mapping.

MODELING

Modeling is a physical or graphical representation of a concept or object. A model, by its very nature, is incomplete; it is only a high level abstraction. A prototype is an example of a model. A model often consists of basic components and their relationships among one another. Also, the relationships are reflected via the exchange of signals, data, or both.

GOALS

- Identify concepts or objects.
- Identify their relationships.
- Simply the explanation of relationships.
- Determine the impact of a change to concepts or objects and their relationships.

OBSTACLES

- Assuming a model is complete
- Believing that assumptions behind a model are facts
- Not providing a legend to the model diagram
- Developing a model that is too complex or unclear
- Assuming that all components and relationships are equal

Steps

1. Determine:
   a. Purpose of the model
   b. Type of model (e.g., graphical, mathematical)
   c. Components and sources (e.g., data, time) of their relationships
   d. Level of detail
2. Define the rules for its execution.
3. Refine the model, as required, to reflect reality.
4. Provide a legend.

MULTIPLE INTELLIGENCES

The theory of multiple intelligences, developed by Dr. Howard Gardner, is that seven intelligences exist that a person develops or demonstrates over time: *linguistic*, for words; *spatial*, for images; *logical* and *mathematical*, for logic and numbers, respectively; *interpersonal*, for dealing with other people; *intrapersonal*, for self-awareness; and *musical*, for rhythm and sounds. Traditionally, intelligence was measured by pencil and paper tests, which were measures of only one kind of intelligence. Only a few people develop and exhibit more than two of the seven intelligences. The manifestation of an intelligence depends to a large degree on how receptive an environment is to the development and expression of that intelligence.

Goals

- Identify the different types of intelligences available among stakeholders.
- Apply people in a manner that capitalizes on specific intelligences of individuals.
- Encourage solid teaming arrangements that require the input of different intelligences.

Obstacles

- Not understanding the concept of multiple intelligences
- Assuming that people generally have only one category of intelligence

Steps

1. Determine those categories of intelligence that you do or do not exhibit.
2. Observe the behavior of others on your team to determine those categories of intelligence that they do or do not exhibit.
3. Use this knowledge to assign tasks and to determine teaming relationships that are conducive to the category of intelligence for individuals.

MULTIVOTING

Multivoting is a formalized voting procedure where participants silently vote on the best ideas or options. Each person has a limited number of votes to cast, often of different value. This technique is best used when a number of items must be selected. Once all the votes are taken, they are totaled based upon the value of each vote. Naturally, the items with the highest scores are selected.

GOALS

- Provide an equitable way to resolve disputes.
- Pursue a more objective way to select among various items or options.

OBSTACLES

- Being unclear on the voting process
- Not including the appropriate people during the vote

STEPS

1. Define the issue to be voted on.
2. Determine who can participate.
3. Seek agreement or consensus on the rules for multivoting.
4. Define the acceptable result.
5. Assign someone to tally the votes.
6. Honor the result.

MYERS–BRIGGS TYPE INDICATOR

The Myers–Briggs Type Indicator (MBTI) is a psychological approach to determine character and temperament types. The sixteen types are based upon four dichotomous pairs of preferences. *Extraversion* vs. *introversion* reflects whether a person has a preference, for example, for sociability or territoriality. *Intuition* vs. *sensation* reflects whether a person has a preference, for example, for imagination or practicality. *Thinking* vs. *feeling* reflects whether a person has a preference, for example, for logic or emotional sensitivity. *Judging* vs. *perceiving* reflects whether a person has a preference, for example, for finality or open-end scenarios.

GOALS

- Identify the preferences of stakeholders.
- Apply people in a manner that reduces negative conflict.
- Encourage solid teaming arrangements.

OBSTACLES

- Not understanding and applying all the principles behind MBTI
- Assuming that people generally fall rigidly into one of the types

Steps

1. Determine your personality type using the four categories of character and temperament types.
2. Observe the behavior of others on your team from the perspective of the four categories of character and temperament types.
3. Use this knowledge to assign tasks and to determine teaming relationships that are conducive to the personality types of the individuals.
4. Determine the positive and negative impacts of your character and temperament type.

N

NEGOTIATING

Contrary to popular belief, a successful negotiation session is one where the result is a win–win result for all parties. That is, all parties leave feeling positive about the results. To achieve a win–win result does not come easy. It requires having an understanding of your wants and needs and that of the other party. It requires then developing an effective strategy that is executed using a selected set of tactics. A successful negotiation does not mean rigidly sticking to a game plan. Instead, it requires the flexibility to adapt based on continual feedback. If need be, it consists of a willingness to walk.

Goals

- Achieve a win–win result.
- Achieve goals and objectives.

Obstacles

- Seeking a win–lose, lose–lose, or lose–win result
- Failing to understand the relationship between strategy and tactics
- Failing to determine needs and wants of oneself and the other party

Steps

1. Determine the overall goals and objectives you hope to achieve during negotiation.
2. Develop a strategy to achieve them.
3. Determine the tactics to employ.
4. Try to understand as much as you can about the other party, such as goals and the paradigm that he or she might be following.
5. Try to understand as much as you can about yourself, such as assumptions, preferences, and paradigm used.
6. Maintain constant feedback on progress.
7. Seek a win–win outcome.
8. Be willing to walk away from an agreement.

NET PRESENT VALUE

Net present value, also known as discount cash value, is used to determine the best investment among two or more alternatives. The idea is to determine which alternative has the greatest cash inflows *vs.* cash outflows. The key decision is whether the rate of return from investment exceeds that of the interest paid to a bank for borrowing the money. The variables to consider are the cost of the proposed project, the interest charged for borrowing the money, the anticipated savings, and the number of years of the savings. The formula is:

$$\text{Net present value} = [\text{anticipated savings for year } 1/(1 + I)] \\ + \ldots + [\text{anticipated savings for year } n/(1 + I)]$$

where I is the interest rate, and n is the number of years. For example,

$$\$54{,}545 = [\$30{,}000/(1 + .10)] + [\$20{,}000/(1 + .10)] + [\$10{,}000/(1 + .10)]$$

Goals

- Determine whether the present cash flows exceed outflows.
- Determine which project to invest money in.

Obstacles

- Failing to recognize that NPV is only one method of analysis
- Using exaggerated or inaccurate data

Steps

1. Identify the following for each calculation:
   a. Cost of project
   b. Savings due to new product or service
   c. Interest rate
   d. Number of years of savings
2. Calculate the NPV for each scenario (e.g., different project).
3. Chart the results of the calculations.
4. Select the desired option.

NETWORK DIAGRAM

A network diagram is a portrayal of the logical sequence of tasks that identifies their start and stop dates and indicates which ones are critical for the project from a time perspective (Figures 28 and 29). The two types of network diagrams are arrow and precedence diagrams. Precedence diagrams have gained wider acceptance; they use boxes (containing data such as duration, title, unique numeric designator, or dates) to reflect tasks and vectors to show sequence. The network diagram is based upon certain information: the work package level items contained in the work breakdown

Network Diagram

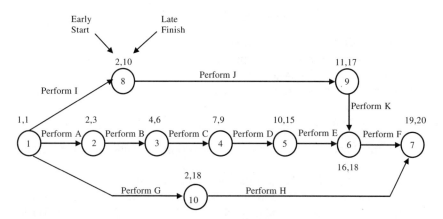

FIGURE 28 Arrow diagram. (From Project Management Seminar presented by Practical Creative Solutions, Inc., 1996.)

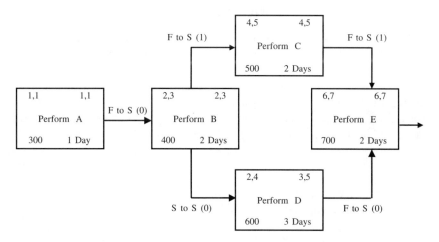

FIGURE 29 Precedence diagram. (From Project Management Seminar presented by Practical Creative Solutions, Inc., 1996.)

structure; the time estimate to perform each task; the logical sequence of the tasks; the type of dependency relationship (e.g., finish-to-start relationship), and lag value. The use of the information results identifying early and late start and stop dates as well as the critical path.

Goals

- Provide a path to reach goals or objectives.
- Maintain focus on a goal or objective.
- Enable better communications.
- Provide an effective reporting tool.

Obstacles

- Not identifying all dependencies
- Drawing the diagram at too high a level
- Assuming all tasks are linear
- Providing diagram to the wrong audience

Steps

1. Use the lowest level (work package level) in the work breakdown structure.
2. Connect the tasks to reflect their logical sequence.
3. Calculate the early start and finish dates via the forward pass for each task.
4. Calculate the late start and finish dates via the backward pass for each task.
5. Calculate the total float for each task.
6. Identify the critical path.

NEURAL NETS

A neural net is a modeling approach based upon the working of the human brain (Figure 30). Like the brain, a neural network is an elaborate array of interconnected cells, or nodes. Each node functions as a switch, such as on or off, strong or weak. A series of nodes are activated to reflect a specific pattern. The qualitative aspects of a signal determine the path that a signal travels through a series of nodes and, consequently, a specific pattern of behavior. The key is to adjust input and output nodes based upon the signal received and to assign a value to a node.

Goals

- Identify the relationships among various nodes.
- Reduce the complexity behind those relationships.

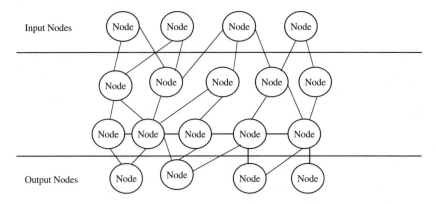

FIGURE 30 Neural net.

Nominal Group Technique

- Identify patterns of behavior.
- Identify how input signals are converted into outputs.

OBSTACLES

- Failing to identify all the nodes
- Failing to identify all the relationships
- Assuming that a spurious relationship is correct

STEPS

1. Determine:
   a. Problem categories (e.g., signals)
   b. Input nodes
   c. Intermediate nodes
   d. Output nodes
2. Connect the nodes to reflect the different kinds of relationships reflected in the form of patterns.
3. Assign weights to the different nodes to reflect different relationships.

NOMINAL GROUP TECHNIQUE

The nominal group technique (NGT) is a structured approach to brainstorming where ideas are combined and eliminated and then the participants select the best option. The idea is to avoid up-front prejudicial judgment that can result in the elimination of good ideas. This technique also provides an organized, logical approach to selecting the best idea.

GOALS

- Reduce the influence of prejudicial judgment.
- Allow unpopular but effective ideas to be considered.
- Provide an organized way to make group decisions.

OBSTACLES

- Allowing groupthink to influence results
- Failing to eliminate ideas or options objectively
- Trying to rush through the NGT process

STEPS

1. Dispense with any preconceived notions.
2. Brainstorm ideas.
3. Determine the criteria for evaluating and dealing with ideas.
4. Apply the criteria.
5. Determine the approach for selecting the best ideas (e.g., multivoting).

O

OBJECT AND PROCESS MODELS

System models often fall into one of two categories, although numerous hybrids also exist. *Process models*, the first type, view the world as a set of functions that relate to one another through inputs, processing, outputs, and movement of data. These models are often used to capture requirements. Some common techniques for process modeling are those of Gane and Sarson and DeMarco. The basic ingredients for process models are functions (or processes), data, decomposition levels, and stores. *Object models* view the world as a set of objects or entities (that is, instances of a class) and the relationships among them. The basic ingredients of object models are objects, relationships, attributes, and states. These models are often used to build designs. Some common techniques for object modeling are those of Coad and Yourdon. Regardless of the category of modeling, use a consistent set of symbolic notations and, once completed, maintain configuration control.

GOALS

- For object models:
  - Identify major components of a system (e.g., product).
  - Identify relationships among the components.
  - Assess the quality of the relationships.
- For process models:
  - Identify major processes of a system (e.g., product).
  - Identify the inputs and outputs of data and other resources for each process.
  - Assess the effectiveness of the processes.

OBSTACLES

- For object models:
  - Not identifying all the components or relationships
  - Failing to identify all the significant characteristics of each component and relationship
  - Not using the model for other models (e.g., entity relationship model) in the future
- For process models:
  - Not identifying all the processes and their respective inputs and outputs
  - Failing to explode processes down to a meaningful level of granularity
  - Not using the model for other models developed in the future

STEPS

1. Determine the most appropriate modeling approach.
2. Determine the symbolic notation to use.
3. Determine the purpose of the model for later activities in the project life cycle.
4. Once it is completed and agreed upon, place the model under configuration control.

OBJECTIVES

An objective is a measurable criterion or criteria supporting the attainment of one or more goals (Figure 31). To be meaningful, an objective should be specific and measurable. An example of an objective is "The project will finish under $500,000." An objective should be one simple statement that contains something measurable.

GOALS

- Give a team a sense of direction.
- Efficiently and effectively execute a project.

OBSTACLES

- Not defining objectives adequately
- Not generating commitment to objectives
- Not recognizing the relationship between goals and objectives

STEPS

1. Determine goals.
2. For each goal, determine one or more objectives that are specific and measurable.
3. Ensure that each objective correlates to one or more deliverables in a work breakdown structure.

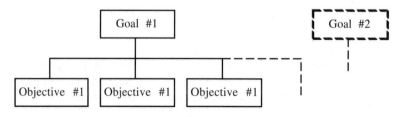

FIGURE 31 Objectives and their relationship to goals.

ORGANIZATION CHART

An organization chart is a graphical display of the reporting relationships and roles on a project (Figure 32). Specifically, it should show the chain of command, portray direct and indirect relationships, and reflect the weighted core of expertise (e.g., software engineering on a software development project). A quick review of an organization chart should indicate the degree of application of the unity-of-command principle and the span of control considerations.

GOALS

- Identify roles and responsibilities.
- Identify reporting relationships.
- Assess adequacy of span of control.

OBSTACLES

- Not showing the key stakeholders
- Not reflecting the weight of expertise required on a project
- Not observing the unity-of-command principle
- Reflecting too many layers of oversight (e.g., too many team leads)

STEPS

1. Identify all the applicable stakeholders.
2. Determine their roles, responsibilities, and relationships to one another.
3. Draft the chart, noting direct and indirect relationships.

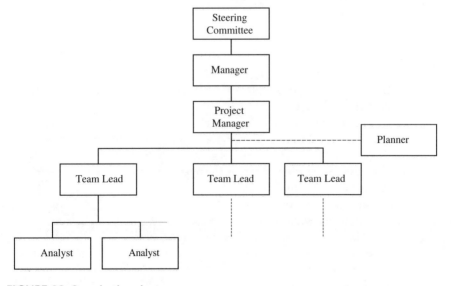

FIGURE 32 Organization chart.

4. Publish the chart.
5. Remember to update it periodically.

ORGANIZATIONAL ENGINEERING

Organizational engineering (OE) is a branch of knowledge that seeks to understand, measure, predict, and guide the behavior of groups of people exhibited over time in response to situations. Behavior is exhibited in two ways: method and mode. *Method* is a person's approach to handling issues, whether structured or spontaneous. *Mode* is a person's approach to responding to information, whether it is immediately or after a length of time. The combination of the two forms creates four primary styles: reactive stimulator, relational innovator, hypothetical analyzer, and logical processor.

GOALS

- Identify the strengths and weaknesses of the information processing styles of stakeholders.
- Utilize people in a manner that capitalizes on their strengths and compensates for their weaknesses.
- Encourage a solid teaming arrangement.

OBSTACLES

- Not applying all the principles behind OE
- Assuming that people always adhere to one of the primary styles

STEPS

1. Steps for implementing OE on a new team:
   a. Do an assessment of yourself.
   b. Determine the tasks and the nature of the work to be performed for a project.
   c. Determine the desired style for each task.
   d. Match the styles of people with the type of tasks they are to perform.
   e. Monitor performance.
   f. If necessary, reassign or redirect people.
2. Steps for managing an existing team:
   a. Do an assessment of yourself.
   b. Determine the status of existing tasks.
   c. Conduct an inventory of the styles on your team.
   d. Determine the style characteristics required for the remaining tasks.
   e. Match the styles of individuals with the style requirements of the tasks.
   f. If necessary, reassign people.
   g. Monitor performance and take corrective action, if necessary.

ORGANIZING

Organizing, one of the four major functions of project management, orchestrates resources efficiently and effectively to execute project plans. It includes activities such as organizing a team; preparing procedures and workflows; preparing forms, reports, and memos; and establishing a project library, project history files, and project manuals.

Goals

- Employ resources efficiently and effectively.
- Reduce or eliminate negative conflict among team members.

Obstacles

- Treating organization as a secondary consideration relative to other functions
- Emphasizing efficiency over effectiveness
- Providing inadequate resource availability
- Not clarifying roles and responsibilities
- Using poor judgment, as revealed by applying too little or too much organization

Steps

1. Determine and procure the necessary resources.
2. Determine the support infrastructure to have in place.
3. Ensure that organizing activities support achievement of the vision, goals, and objectives.
4. Integrate with the planning, controlling, and leading functions of project management.

OUTSOURCING

Outsourcing is contracting with another company to provide services that would ordinarily be provided in-house. The reasons for outsourcing could be cost, skill shortage, or a desire to keep overhead low. Whatever the reason, it is important to determine the overall goals and objectives of an agreement and the accompanying strategy to achieve them. Ideally, the agreement should be as clear as possible to both the outsourcer and the team members. One of the most overlooked consequences of outsourcing is its effect on employees. Often, it can have a high negative impact, particularly if employees feel threatened. In addition, other risks might include issues related to security, legal liability, intellectual capital, and vague terms and conditions.

Goals

- Obtain the best service for the lowest price.
- Maintain uninterrupted service.

- Ensure that service standards are sustained.
- Provide effective monitoring of service-level agreements.

OBSTACLES

- Failing to monitor performance of outsourcing services
- Incorporating vague terms and conditions in the outsourcing agreement
- Not assessing the impact and risks on a project team

STEPS

1. Identify core competences.
2. Determine the services to outsource.
3. Determine the type of outsourcing agreement desired.
4. Perform a cost/benefit analysis.
5. Conduct an impact analysis, looking not just at impacts on technological and operational concerns but also the impact on people.
6. Select potential outsourcing firms.
7. Conduct negotiations, keeping in mind goals and strategies.
8. Monitor performance against the contract.

P

P²M²

The practical project management methodology, or P²M², is a nonlinear, integrated approach for managing projects. What distinguishes it from other methodologies is its emphasis on the role of people in a project. The idea is that if people are managed effectively then cost, schedule, and quality criteria will be satisfied. The major components behind P²M² are customer satisfaction, balance, self-management, modularity, process improvement, nonlinearity, metrics, integration, simplicity, motivation, flatness, adaptability, divergence, and effectiveness. The iterative cycle of P²M² consists of these five elements: inputs, tasks, responsibilities, outputs, and measures of success.

Goals

- Plan, organize, control, and lead projects efficiently and effectively.
- Recognize that a project is people, not task, oriented.

Obstacles

- Not understanding the basic concepts behind the methodology
- Becoming inflexible in its application
- Applying traditional project management approaches under the guise of P²M²

Steps

1. Recognize that project management is a nonlinear discipline.
2. Understand that people are the ingredients for successfully completing a project.
3. Understand that the basic functions of project management require taking an integrated approach.
4. Apply the P²M² cyclic approach when planning and managing a project.

PARADIGM

A paradigm is a mental model or framework for understanding, interpreting, and dealing with reality. It consists of a body of assumptions, beliefs, values, theories, knowledge, etc. It helps people to make choices and determine ways to respond to reality consistent with the prevailing paradigm.

GOALS

- Provide an orderly way to acquire and process data and information.
- Provide structure to assess and deal with circumstances.

OBSTACLES

- Assuming that the paradigm is the "truth"
- Casting aside any important information that does not justify the paradigm
- Attempting to fix a paradigm that clearly is based upon erroneous information and assumptions
- Concentrating on addressing the needs of the paradigm rather than solving a problem or overcoming an obstacle

STEPS

1. Determine:
   a. What paradigms exist
   b. The prevailing paradigm
   c. Impacts of the prevailing paradigm
2. For positive impacts, determine how to maximize the benefits.
3. For negative impacts, determine how to mitigate their effects or avoid them.
4. Ascertain the elements and rules of the prevailing paradigm.

PARETO ANALYSIS CHART

A graphical tool is used to distinguish the major causes of problems from minor ones (Figure 33). The principle is based on the notion that 20% of the causes contribute to 80% of the effects. In other words, the few of anything contribute to achieving the most. One axis of the chart has gradations of percentages up to 100%, and the other axis reflects the cumulative number of individual observations in a particular category.

GOALS

- Identify causes.
- Perform unbiased, objective analysis.

OBSTACLES

- Insufficient observations or occurrences
- Insufficient objective criteria for selecting observations or occurrences

STEPS

1. Recognize that 80% of the effects are created by 20% of the causes.
2. List all the characteristics of each category of observation.
3. Determine the sufficient number of observations.

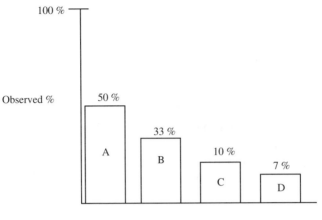

FIGURE 33 Pareto analysis chart.

4. Conduct, record, compile, and analyze observations.
5. Build the Pareto analysis chart.
6. Apply statistical tests to determine the cause of the largest percent of observation.
7. Take corrective action to reduce the size of the most observed occurrences.
8. Take new observations to determine that fixing the cause has reduced the highest percentage of results.

PARKINSON'S LAW

Parkinson's law is simple in concept: a task will take the time available for its completion. For example, if task A is given 10 hours for completion, then it will take that long. If it is given 20 hours, then it will take that long. Often, the time available for completion is either too much or too little. While a task may be completed within a given time, too short of an assigned time can result in a host of dysfunctional, nonproductive behaviors, such as short cuts that have immediate payback but result in long-term liabilities. It is important, therefore, to provide a realistic time frame to ensure a greater likelihood of effective and efficient productivity. The way to achieve that is by using good estimating techniques, such as the Program Evaluation and Review Technique (PERT) formula.

GOALS

- Provide adequate time to complete a task.
- Recognize that the time to do a task is frequently a matter of perception.

OBSTACLE

- Estimating too much or too little time to perform a task

STEPS

1. Use estimating techniques, such as the PERT formula.
2. Determine the degree of inaccuracy that an estimate might contain.
3. Make adjustments accordingly.

PDCA CYCLE

The PDCA (plan, do, check, act) cycle, also known as the Deming wheel, is a tool for quality improvement that can also be used as a decision-making tool (Figure 34). *Plan* is determining what to achieve and the respective tasks. *Do* is executing the plan. *Check* is measuring performance. *Action* is analyzing findings and making improvements. The PDCA cycle is iterative.

GOALS

- Improve quality decision making and performance.
- Reduce variability in outcomes.

OBSTACLES

- Failing to repeat the cycle on a continual basis
- Failing to perform *check* and *act* consistently

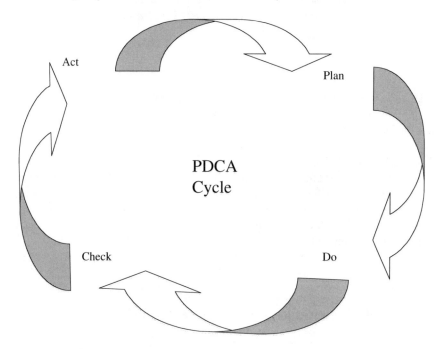

FIGURE 34 PDCA cycle.

STEPS

1. Follow the steps in the PDCA cycle.
2. Determine what steps to perform under:
   a. Plan
   b. Do
   c. Check
   d. Act
3. Apply the PDCA cycle when dealing with issues related to quality or when making decisions.

PEAK EXPERIENCE

Peak experience is a deep emotional involvement with a task. A person who has a peak experience is someone who expends extensive time and effort in a task but does not realize it. Time seems to fly, as other concerns do not seem to interfere with the involvement of that person, who becomes "one" with the task. Abraham Maslow referred to this phenomenon as peak experience, similar to what creative writers refer to as the continuous dream. The psychologist M. Csikszentmihalyi discusses something equivalent to the peak experience as flow. Whether it is peak experience or flow, the idea is to provide people with the opportunity to attain high levels of satisfaction in whatever they do on a project.

GOALS

- Increase an individual's efficiency and effectiveness.
- Increase enthusiasm and morale.
- Generate commitment and a sense of responsibility.

OBSTACLES

- Not appreciating the immense impact of the peak experience to individual and team productivity
- Managing in a manner that destroys any chance for anyone to have a peak experience

STEPS

1. Encourage people to assume tasks or responsibilities that they can and like to do.
2. Provide the support that enables them to have the peak experience.
3. Recognize that the project manager can only provide the opportunity for people to have peak experiences and cannot force someone to have one.

PERT ESTIMATES

PERT (Program Evaluation and Review Technique) is a scheduling technique that incorporates the use of three estimates to derive the time required to complete a task (Figure 35). Over the years, the PERT scheduling technique and critical path method (CPM) have merged to develop network diagrams incorporating the best of both. The PERT estimating technique, also known as the three-point estimate, requires making three estimates: most optimistic, most likely, and most pessimistic. The estimates are then plugged into the following formula:

$$\{\text{most optimistic} + [4 (\text{most likely})] + \text{most pessimistic}\}/6$$

The *most likely* time is the required time to complete a task under normal or reasonable conditions. The *most pessimistic* time is the time required to complete a task under the worst or nightmarish conditions. The *most optimistic* time is the time required to complete a task under the best or ideal conditions. The formula results in an expected time to do the work of a task. It is does not, however, account for nonproductive time (e.g., disruptions). The expected time is then adjusted to reflect the effect of nonproductive time. The result is a 99% confidence level that a task will fall within the revised expected time.

GOALS

- Give reasonable assurance that estimates are reliable.
- Reduce the level of subjectivity in estimates.
- Discuss early on some of the major considerations to execute tasks.

Task	Most Optimistic	Most Likely	Most Pessimistic	Expected Time
Perform A	8	32	50	31.0
Perform B	2	8	24	9.7
Perform C	2	8	12	7.7
Perform D	16	40	80	42.7
Perform E	2	10	14	9.3
Perform E	1	3	7	3.3
Total	31	101	187	103.7

FIGURE 35 PERT estimating technique. (From Project Management Seminar presented by Practical Creative Solutions, Inc., 1996.)

Obstacles

- Not performing PERT calculations consistently
- Skewing calculations to what one expects

Steps

1. Solicit estimates from relevant stakeholders for each task at the lowest level (work package level) of the work breakdown structure.
2. For each task, determine the most likely, most optimistic, and most pessimistic times.
3. Adjust estimates to reflect nonproductive time.
4. Convert the revised expected time to duration, if necessary.
5. Use estimates to calculate budget.
6. Use estimates to determine schedule dates.
7. Review estimates with stakeholders.
8. Revise estimates.

PETER PRINCIPLE

The Peter principle, a well-known management rule that has been accepted for a long time, states that eventually a person will reach his highest level of incompetence within an organization. The Peter principle appears on all levels of the corporate hierarchy. It also applies to projects, because people can reach levels of responsibilities for one or more tasks that they are truly unable to perform. Often, however, this inability to perform results from the Pygmalion effect rather than incompetence. The Pygmalion effect is that a person will become what is expected of him. If the person is expected to be a pro, then that person will become a pro. If that person is expected to be incompetent, the expectation will likely be realized. Be aware, however, that a lack of job enrichment, job challenge, or opportunities can be misinterpreted as being due to the Peter principle.

Goals

- Provide better opportunities for growth.
- Recognize when people fail to perform according to requirements or expectations or both.

Obstacles

- Not recognizing when the Peter principle occurs
- Not doing anything when the Peter principle occurs
- Misdiagnosing the Peter principle as being the cause when actually the Pygmalion effect has come into play

STEPS

1. Learn as much as possible about the skills, knowledge, etc. of an individual.
2. Establish and follow up on standards of performance.
3. Try to ascertain the cause for substandard performance.
4. Take corrective action and follow up accordingly, if necessary.

PLANNING

Planning is one of the four functions of project management. It involves deciding in advance what a project will achieve, determining the required steps, and identifying when it will be completed. Planning entails determining the goals and objectives via a statement of work, developing work breakdown structures, performing time and cost estimating, building schedules, and allocating resources. Planning is very difficult because it requires clarifying many ambiguities and assumptions up front, conducting extensive negotiations, engaging the involvement of key stakeholders, and competing with other projects. With a solid plan, however, a project can progress much more smoothly throughout its life cycle.

GOALS

- Provide a road map to achieve goals and objectives.
- Encourage stakeholder confidence to achieve goals and objectives.

OBSTACLES

- Not providing sufficient time
- Not providing planning expertise
- Not having access to important information
- Not having cooperation from stakeholders
- Not providing sufficient resources

STEPS

1. Prepare a statement of work or statement of understanding.
2. Develop a work breakdown structure.
3. Develop cost and time estimates.
4. Generate a schedule.
5. Allocate resources.
6. Obtain consensus from all stakeholders after developing the plans.

POST-IMPLEMENTATION REVIEW

A post-implementation review is an objective determination of whether a project has met its goals and objectives. Disinterested parties conduct a review using specified criteria to assess performance throughout the life cycle of a project. The result is a report that contains findings and recommendations for improvement. Ideally, it is best to let some time lapse between the completion of a project and the beginning of its review. Time allows issues to settle and resolve themselves. A major contribution of a review is that it enables fixing discrepancies and provides insights for similar projects in the future.

GOALS

- Verify that goals and objectives of project have been achieved.
- Verify that requirements and specifications have been met.
- Determine if any corrective action is necessary.

OBSTACLES

- Not using objective criteria
- Failing to use a disinterested person
- Not following up on findings and recommendations
- Not documenting results

STEPS

1. Determine who should conduct the review.
2. Determine the most appropriate time to conduct the review.
3. Document the results.
4. Assess the results.
5. Develop a corrective action plan, if necessary.
6. Implement the plan.
7. Collect feedback on the results of corrective actions.

POWER

Power, defined as the capacity to achieve specific results, comes in many forms. John French and Bertram Raven identified five bases of power: *Coercive power* is getting people to achieve results through force or fear. *Reward power* is using positive tangible and intangible incentives. *Legitimate power* stems from one's position in an organizational hierarchy. *Expert power* is having an extraordinary skill, expertise, or knowledge over others. *Referent power* relies upon the personal characteristics of the leader (e.g., charisma). Most project managers rely upon legitimate, expert, and referent power, although in some circumstances they can also exert coercive and reward power.

GOALS

- Identify the appropriate type of power available to achieve results.
- Recognize that power comes in many different forms and can be exercised differently in different circumstances.
- Recognize the limits of one's power to achieve results.

OBSTACLES

- Failing to recognize one's source of power
- Exercising one's power inappropriately
- Assuming that one has no power

STEPS

1. Identify the formal and informal powers at your disposal.
2. Ascertain whether you are applying power effectively.
3. Avoid the use of coercive power due to its long-term negative effects.
4. Recognize that project managers basically have expert and referent powers.
5. Recognize, however, a project manager can exert the other three powers by influencing stakeholders who possess those other powers.

PRESENTATION

A presentation fundamentally has one of three purposes (which can be amalgamated in various combinations): persuade, inform, or explain. All presentations should have three sections: introduction, main body, and conclusion. Each presentation should go through the following stages: perspective, perception, planning, preparation, practice, and performance.

GOALS

- Ensure better communications.
- Reduce negative conflict and misunderstandings.

OBSTACLES

- Using too much jargon
- Not considering the needs of the audience
- Using unclear, unfocused content
- Providing poor presentation

STEPS

1. Determine perspectives.
2. Determine perceptions.
3. Plan the presentation.
4. Prepare the presentation.
5. Practice giving the presentation.
6. Deliver the presentation.

PRESENTATION: PERCEPTION

Perception is the stage of giving effective presentations that addresses how you perceive your audience and how it perceives you. It involves building positive relationships with the audience, maintaining a good appearance and poise, demonstrating expertise and knowledge, and understanding any differences.

GOALS

- Ensure better communications.
- Reduce conflict and misunderstandings.

OBSTACLES

- Not understanding the characteristics of an audience
- Not understanding the necessary values, beliefs, etc. to relate to an audience

STEPS

1. Determine what you, as the speaker, have in common with the audience.
2. Determine what you, as the speaker, do not have in common with the audience.

PRESENTATION: PERFORMANCE

Performance is the stage of project performance that addresses delivery of the presentation. It involves using vocal variety, maintaining eye contact, using good body movement, obtaining audience involvement, and showing enthusiasm.

GOALS

- Ensure better communications.
- Reduce negative conflict and misunderstandings.

OBSTACLES

- Coming across as too technical or too general
- Coming across as defensive or too condescending
- Not relating to an audience

STEPS

1. Maintain good eye contact.
2. Move about the podium.
3. Avoid distracting mannerisms.
4. Encourage audience involvement.

PRESENTATION: PERSPECTIVE

Perspective is the stage of giving an effective presentation that involves knowing yourself and your audience. As a speaker, you should perform an audience analysis (e.g., physical characteristics and values) and an analysis of yourself (e.g., physical characteristics and values). The result is to come to a win–win answer to the question of "what's in it for me?"

GOALS

- Ensure better communications.
- Reduce negative conflict and misunderstandings.

OBSTACLES

- Not focusing the material on the needs of the audience
- Developing a too-detailed or too-general presentation

STEPS

1. Determine the audience.
2. Determine the purpose of the presentation.

PRESENTATION: PLANNING

Planning is the stage of giving an effective presentation when the speaker determines its type and structure. Specifically, planning is deciding on the overall purpose of a presentation; developing the introduction, main body, and conclusion; and answering the five basic questions of who, what, when, where, and why.

GOALS

- Ensure better communications.
- Reduce conflict and misunderstandings.

OBSTACLES

- Not providing structure to a presentation
- Not using an appropriate organizational structure for a topic

STEPS

1. Determine where and when the presentation will be given.
2. Determine the length of the presentation.
3. Determine the overall structure of a presentation.
4. Determine the physical requirements (e.g., arrangement of the room).
5. Determine necessary supplies.

PRESENTATION: PRACTICE

Practice is the stage of giving effective presentations when the speaker rehearses to improve performance. It requires trying different modes of practicing, and doing so more than once; concentrating on delivery; and striving for clarity and conciseness.

GOALS

- Ensure better communications.
- Reduce negative conflict and misunderstandings.

OBSTACLES

- Not practicing realistically
- Not receiving adequate feedback

STEPS

1. Practice in an environment similar to actual presentation.
2. Practice being natural.
3. Practice, if necessary, in front of peers to acquire valuable feedback.
4. Identify what can go wrong and take corrective action, if necessary.

PRESENTATION: PREPARATION

Preparation is the stage of giving effective presentations in which the material itself is developed. It requires applying logic and emotion, using supporting materials, keeping visual aids simple, and selecting the mode of delivery.

GOALS

- Ensure better communications.
- Reduce negative conflict and misunderstandings.

OBSTACLES

- Not doing thorough research
- Inadequately preparing the presentation

STEPS

1. Conduct research on a topic.
3. Determine the logical progression of material (e.g., chronological).
3. Ensure clarity of material.
4. Use a variety of presentation techniques (e.g., pictures, bullet lists, etc.).

PRIORITIES OF CHANGE

Not all changes are equally important. If they were, a project would be constantly reacting to them. Some changes are classified as a "major priority" and must be addressed right away. Often, these changes, also known as showstoppers, bring a project to a complete halt if not addressed. Some are a "minor priority" and do not require immediate attention but must be addressed before completing a project. Some are "deferred changes" and can be addressed when time permits, even after the project is complete.

Goals

- Provide better management of incoming changes.
- Enable better allocation of resources.

Obstacles

- Inability to come to a definition of each priority
- Failure to define and follow a response to a particular priority of a change

Steps

1. Develop medium (e.g., form) to capture information regarding a change.
2. Develop criteria to prioritize a change.
3. Apply criteria.
4. Notify person who requested the change, if necessary.

PROBABILITY

Probability is drawing inferences, or predictions, from data and then testing for the likelihood of occurrence. Two types of probability exist: subjective and objective. *Subjective probability* reflects judgment on the part of the individual, based upon, for example, experience and knowledge. *Objective probability* reflects the use of statistical observation having no predetermined preferences. Probability is the likelihood of occurrence of a set of events according to some sequence, or permutation. An event is a discrete outcome. An event can be simple or compound. A simple event is an event that cannot be subdivided; a compound event consists of two or more simple events. Probability, then, involves determining the likelihood of occurrence of one or more events that are either exclusive or conditional. An exclusive event is not dependent upon the occurrence of an earlier event; a conditional event is one in which a dependency does, in fact, exist.

Goals

- Observe relationships among data and their associated variables.
- Conduct inference from data collected.

Obstacles

- Not identifying all the key dependent and independent variables
- Assuming that a spurious relationship exists between two or more variables
- Relying upon biased data or an inadequate sample size

STEPS

1. Obtain a meaningful sample of data.
2. Identify dependent and independent variables.
3. Perform probability calculations.
4. Draw inferences from the sample.
5. Conduct additional tests.
6. Compare inferences with actual test results to draw conclusions.

PROBLEM SOLVING

The key for successfully solving a problem is first to define it thoroughly. Too often, people jump to a solution before really understanding what they need to fix. The result is often an unworkable solution that only leads to greater problems later. It is important, therefore, to answer the who, what, when, where, why, and how of a problem. Satisfied with that information, a number of alternatives can be developed that can be narrowed down to the best one. The selection is then implemented and regular feedback is taken to determine any necessary adjustments. One of the major difficulties of following this approach is that people fail to question their hidden assumptions, not realizing that it affects their selection of a solution. It is important, therefore, to take different perspectives by thinking outside the box, inside the box, inside-out of the box, and outside-in of the box.

GOALS

- Clearly define a problem before choosing and implementing a solution.
- Determine the best solution.
- Encourage innovative thinking.
- Rely on facts and data, not assumptions and guesses.

OBSTACLES

- Jumping to the solution before defining the problem
- Introducing biases in the form of preferences to determine a solution
- Failing to identify and address alternatives

STEPS

1. Define a problem as thoroughly and specifically as possible, including the scope.
2. Identify the who, what, when, where, why, and how of a problem.
3. Develop alternative solutions.
4. When developing alternatives, use creative thinking and identify any prejudices or hidden assumptions.
5. Select the best solution through comparative analysis involving strengths and weaknesses of each one.

6. Develop a plan.
7. Implement the plan.
8. Seek feedback on effectiveness.
9. Take corrective action, if necessary.

PROCEDURES

Procedures are documents that describe the major activities for executing the four functions of project management. Procedures can cover a wide arrange of topics, including schedules, change control, meetings, responsibilities, organizational structure, equipment utilization, supply purchases, and forms completion. Procedures often take one of four formats: item-by-item, narrative, sequential, and playscript. Often, procedures are supplemented with work flows and other diagrams and charts.

GOALS

- Improve communications.
- Enable everyone to work on the same wavelength.
- Improve productivity.

OBSTACLES

- Preparing at a too high a level
- Preparing at a too detailed level
- Using vague, ambiguous language
- Making procedures inaccessible to stakeholders who need them

STEPS

1. Determine what procedures are necessary.
2. Determine what topics should be covered and at what level of detail.
3. Determine who should prepare the procedures.
4. Determine who should get copies of the procedures and how many should be produced.
5. Determine who will maintain them.
6. Determine how often they will be reviewed.

PROJECT

A project is a temporary endeavor to develop a product or deliver a service. More specifically, a project has these characteristics: a start and stop date; vision, goals, and objectives; logical sequence of actions; and a final deliverable. Contrary to popular belief, projects exist in all industries, from manufacturing to pharmaceutical, not just construction, information systems, and engineering.

GOALS

- Be on schedule.
- Stay within the budget.
- Meet requirements and specifications.
- Satisfy the customer.

OBSTACLES

- Sacrificing cost and quality for schedule
- Sacrificing cost and schedule for quality
- Sacrificing schedule and quality for cost
- Experiencing scope creep
- Having no or unclear determinants for success

STEPS

1. Look for any business and technical initiatives and other endeavors that could qualify for being classified as a project.
2. Determine how the record of performance could benefit from project management expertise.
3. Determine what project management disciplines can be applied.
4. Apply the disciplines with the consent of the stakeholders.

PROJECT HISTORY FILES

Project history files serve as repositories for containing information about a project. Typical contents include procedures, drafts of all significant documentation, work breakdown structures, correspondence (e.g., memos) with stakeholders, types and versions of schedules, minutes, completed forms, reports, time and cost estimates, and responsibilities.

GOALS

- Enable good traceability.
- Facilitate audits.
- Protect certain stakeholders.
- Develop reports more easily.
- Reduce learning curve.
- Conveniently store project documentation.
- Spend more time on productive activities.
- Improve communications.

OBSTACLES

- Not maintaining the project history files
- Not assigning responsibilities for maintaining the files

STEPS

1. Determine who will establish and maintain the project history files.
2. Determine the organization of the files.
3. Publish and distribute the files.
4. Select a central location for storing the files.

PROJECT LIBRARY

A project library is a central location in the work area of a project designated to store project documentation. Typical contents of a library include policies, procedures, manuals, newsletters, history files, and literature.

GOALS

- Have a central repository of information.
- Make information accessible.
- Have a source for training materials.

OBSTACLES

- Not keeping the library up to date
- Not providing adequate access and tracking of materials

STEPS

1. Determine the contents of the library.
2. Determine who will establish and maintain the library.
3. Set up the library.
4. Institute mechanisms to track removed materials.

PROJECT LIFE CYCLE

A project life cycle is a logical sequence of phases (Figure 36). The four basic phases are defining, designing, developing, and delivering. These phases occur either linearly or nonlinearly. Each phase has a unique set of tasks and deliverables, a time frame, weighted level of effort in regard to the other phases, and checkpoints. The type of life cycle, such as waterfall or spiral, depends on the context of the environment.

GOALS

- Provide an orderly approach to plan and manage a project.
- Encourage effective and efficient use of resources.

Project Management 129

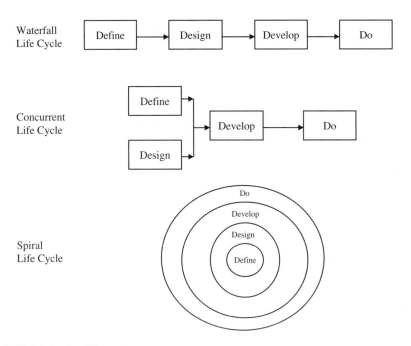

FIGURE 36 Project life cycles.

OBSTACLES

- Getting bogged down in one phase
- Attempting to short-cut certain phases
- Taking a linear approach when a nonlinear one makes more sense or vice versa

STEPS

1. Determine the type (e.g., waterfall, spiral) of life cycle to follow.
2. Determine the phases and their respective deliverables, roles, and responsibilities.
3. Ensure that the infrastructure and history of an organization support the type of project life cycle.
4. Provide the tools to support the type of project life cycle.

PROJECT MANAGEMENT

Project management applies appropriate knowledge, tools, and techniques to plan, organize, control, and lead a project. Good project management techniques further the success of projects through pro-acting rather than reacting, conserving resources, increasing productivity, improving communications, and enhancing quality of workmanship. Projects can avoid some of the causes of project failure, such as a lack of leadership or common vision, failure to learn from mistakes, unrealistic expectations, inadequate planning, lack of solid commitment by stakeholders, overly aggressive schedules, and expanding scope. Project management requires the effective participation of five categories of stakeholders: project manager, senior management, client, project team, and project sponsor. Each one plays an important role during the execution of the four functions of project management (plan, organize, control, lead). The role usually includes one or more of these actions: input, develop, review, and approval.

GOALS

- Build a product or deliver a service that meets or exceeds expectations.
- Increase productivity.
- Enhance communications.
- Gain commitment from stakeholders.
- Ensure effective and efficient resource management.
- Meet commitments dictated by agreements.

OBSTACLES

- Not promoting a sense of direction
- Applying inappropriate tools and techniques for the project
- Not providing an understanding and acceptance of project management
- Viewing project management as an administrative burden
- Not adhering to a schedule
- Avoiding budgetary considerations
- Not accounting for the expectations of stakeholders
- Not considering the people side of project management

STEPS

1. Learn the knowledge, tools, and techniques of project management.
2. Provide the time and other resources to implement project management.
3. Implement project management on a regular basis.
4. Encourage learning about project management by all stakeholders.
5. Use project management throughout the life cycle of projects.

PROJECT MANAGEMENT SOFTWARE

A great variety of software exists for microcomputers, servers, and mainframes. The software can range in capabilities from building a simple bar chart to creating a full network diagram and conducting resource leveling. Ideally, a sophisticated software program should address such capabilities as building a project database, allowing for a large number of activities per project, selecting schedule type, providing specialized and standard reports, developing a baseline and current schedule, entering different logic relationships, assigning one or more resources, generating histograms, performing resource leveling, calculating costs, and conducting what-if scenarios.

Project Management Software

Requirements	Assigned Value	Package #1	Package #2	Package #3
Create a project	2	(5) 10	(5) 10	(5) 10
Duplicate a project	1	(3) 3	(0) 0	(0) 0
Modify a project	2	(5) 10	(3) 6	(0) 0
Handle projects greater than 500 activities	2	(3) 6	(5) 10	(3) 6
Choose between bar charts and network diagrams	1	(0) 0	(2) 2	(0) 0
Provide standard reports	2	(0) 0	(0) 0	(0) 0
Develop customized reports	1	(4) 4	(0) 0	(5) 5
Enter logic relationships	1	(0) 0	(3) 3	(2) 2
Generate resource histograms	2	(4) 8	(0) 0	(0) 0
Perform costing	2	(5) 10	(5) 10	(3) 6
Conduct what-if scenarios	2	(5) 10	(5) 10	(1) 2
Check for network logic errors	2	(5) 10	(3) 6	(0) 0
Perform calendaring	1	(5) 1	(3) 3	(0) 0
Provide import and export capabilities	2	(5) 10	(0) 0	(0) 0
Total Score	80 (max)	82	60	31

Note: Rating is 1-5

Value Legend
2 = necessary requirement
1 = desirable but unnecessary
0 = unnecessary or undesirable

FIGURE 37 Approach for selecting software.

When using project management software, keep in mind that the ability to use the software and apply it to project management concepts requires a learning curve. Requirements for the software should be defined up front (Figure 37).

GOALS

- Improve planning.
- Improve tracking and monitoring activities.
- Increase overall project effectiveness and efficiency.

OBSTACLES

- Not defining requirements thoroughly enough
- Believing cheaper is better
- Confusing project management software expertise with expertise in managing a project
- Failing to check out vendors

STEPS

1. List software and hardware requirements.
2. Give a value to each requirement.
3. Determine which requirements each software package addresses.
4. Tally the points for each package.
5. Select the top-scoring packages for further investigation.
6. Seek out a demonstration of each package, paying attention to value-added, or nonquantifiable, features.
7. Select a package.
8. Train people to use the package.

PROJECT MANAGER

A project manager is responsible for completing a project, as well as planning, organizing, controlling, and leading it. Some of the specific tasks to perform include coordinating effective and efficient participation of team members, developing and managing project plans, enabling communications among stakeholders, and tracking and monitoring project performance.

GOALS

- Complete a project that satisfies cost, schedule, and quality criteria.
- Treat technical, managerial, and people considerations equally.

OBSTACLES

- Not recognizing responsibilities of the role
- Not distinguishing between project leadership and project management
- Ignoring the people side of project management

STEPS

1. Determine formal and informal roles and responsibilities.
2. Determine the context of the environment.
3. Assess formal and informal powers.
4. Determine role expectations.
5. Determine the criteria for selection of project manager.

PROJECT MANUAL

A project manual is a compilation of essential project documentation related to the operational activities of a project. Its content can be just about anything relevant to a project; however, it usually covers such topics as meetings, organizational charts, priorities, procedures, policy statements, resources, reports, forms, change control, schedules, statement of work, personnel listings, and stakeholder relationships and responsibilities.

GOALS

- Provide a central repository of information.
- Make the information accessible.
- Provide a training tool.

OBSTACLES

- Not developing a complete project manual
- Not distributing the manual to appropriate stakeholders
- Not maintaining the manual

STEPS

1. Determine who will plan and organize the project manual.
2. Determine who will receive a copy.
3. Decide who will maintain it.
4. Select the type of binding.
5. Compile the contents.
6. Publish the manual.

PROJECT NEWSLETTER

A project newsletter is a medium to communicate information to all project participants. It can be in hard copy or electronic format (e.g., on the intranet). It can cover such topics as successes, obstacles (even failures), stakeholder biographies, methods and techniques, solutions to common problems, and performance statistics.

GOALS

- Highlight major achievements.
- Share information.
- Facilitate communications.

OBSTACLES

- Becoming enmeshed in politics prior to publication of the newsletter
- Allowing the newsletter to become a propaganda rag
- Publishing the newsletter too late to have any meaning

STEPS

1. Determine who will be responsible for producing the newsletter.
2. Determine the acceptable level of time and effort to devote to it.
3. Assign editorial responsibility.
4. Identify topics.
5. Develop an editorial calendar that indicates topics and publication dates.
6. Identify contributors to each issue.

7. Determine who will review the material.
8. Determine the layout and type of distribution (e.g., website, e-mail, hardcopy).
9. Draft articles or obtain materials.
10. Publish the newsletter.
11. Retain in project history files.

PROJECT OFFICE

A project office is a central location for housing the administrative infrastructure for a medium to large project. Often, the project manager, core team members, and client representatives are located at the project office. The project office is equipped with communications equipment, software, project history files, project library, visibility walls or rooms, meeting and work rooms, etc.

GOALS

- Centrally locate the administrative infrastructure of a project.
- Provide access to project resources.
- Serve as a communications hub for a project.

OBSTACLES

- Failing to provide sufficient resources at the project office
- Allowing the project office to serve as a haven from daily project activities for certain stakeholders
- Allowing the project office to become too much of an overhead cost compared to the cost of a project

STEPS

1. Determine:
   a. Overall purpose of the office
   b. Who will be located at the office
   c. Layout of the office
   d. Resources (e.g., equipment, supplies)
2. Avoid the tendency of the office to become a bunker, isolated from the daily activities of the project.

PROJECT SPONSOR

The project sponsor is the senior manager, or higher, who sponsors a project, appoints the project manager, pays for the project, facilitates the acquisition of resources, and gives the final approval of all key deliverables, changes, and decision requests. In other words, the project sponsor provides guidance and support for a project throughout its life cycle. Often, the project sponsor is on the client's side.

GOALS

- Provide the necessary political muscle for a project to compete with other equally important projects.
- Provide the necessary resources to support a project.

OBSTACLES

- Unclear roles and responsibilities
- Lack of sincere commitment to a project's success
- Lack of political strength

STEPS

1. Determine formal and informal roles and responsibilities.
2. Assess formal and informal powers.
3. Determine context of the environment.
4. Determine role expectations.
5. Determine level of commitment and support, particularly political.

PROJECT TEAM

The project team is composed of the individuals responsible for building the product or delivering the service. Specific responsibilities include providing the necessary expertise and creativity, supporting the project manager, and working directly with the client. Often, a team consists of core team members who are responsible for providing direct support throughout the life cycle of a project. The remaining members provide auxiliary support as required.

GOALS

- Employ resources efficiently and effectively.
- Reduce negative conflict by providing win–win solutions.
- Obtain the necessary people with the requisite skills, knowledge, and experience.

OBSTACLES

- Allowing too much internal negative conflict over roles and responsibilities, resources, and personalities
- Lacking commitment and involvement by members
- Allowing teaming to occur too late down the project life cycle

STEPS

1. Determine formal and informal roles and responsibilities.
2. Assess formal and informal powers.
3. Determine the context of the environment.

4. Determine role expectations.
5. Determine members.
6. Determine criteria to select team members.

PROJECT WALL

A project wall displays important information about a project (Figure 38), including such items as bar charts, network diagrams, organization charts, bullet charts, pictures, and just about anything else deemed important by stakeholders.

GOALS

- Communicate data and information to stakeholders.
- Allow access to data and information.

OBSTACLES

- Incomplete or inaccurate data and information displayed
- Failure to refer to the wall during project execution
- High overhead maintenance of the wall
- Lack of organization of the material on the wall

STEPS

1. Determine:
   a. Purpose of the wall
   b. Audience
   c. Contents

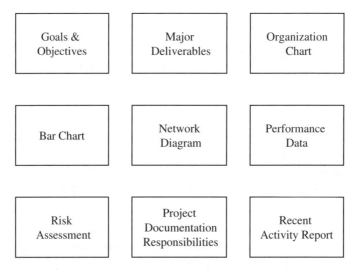

FIGURE 38 Project wall layout.

2. Ensure that the wall is in a location visible to all stakeholders.
  3. Populate the wall.
  4. Maintain the wall by keeping the contents current and accurate.
  5. Use the wall during the conduct of meetings and working sessions.

PROJECT WEBSITE

A project website is a location on a server that provides data and information about a project. Often, it can be a substitute for a project manual for stakeholders who have access to a computer. A website can provide data and information on schedules, cost, and quality as well as administrative matters (e.g., contact lists).

GOALS

- Communicate data and information to stakeholders.
- Allow access to data and information.
- Enable dynamic update of data and information.

OBSTACLES

- Not updating the website
- Not providing all stakeholders with the ability to access the website
- Placing inaccurate or incomplete information on the website

STEPS

1. Determine:
   a. Purpose of the website
   b. Audience
   c. Contents
   d. Level of technical support required
   e. Frequency of updates for contents
2. Populate the website.
3. Use the website during meetings and working sessions.

PROTOTYPING

Prototyping is building a model of a product before it is built on a full scale. Three choices are made regarding a prototype: throw it away, use it as a final product, or enhance it to become the final product.

GOALS

- Reduce product development costs.
- Encourage communications among stakeholders.
- Identify better ideas.
- Identify different ways to satisfy requirements and specifications.

Obstacles

- Not applying prototyping in a meaningful context
- Not matching the prototype with meeting requirements or specifications
- Giving in to the tendency to deploy the prototype into production immediately

Steps

1. Determine the goal of a prototype.
2. Determine the type of prototype.
3. Define requirements and specifications.
4. Build the prototype.
5. Meet with stakeholders to evaluate how well the prototype meets the requirements and specifications.
6. Modify the prototype.
7. Obtain consensus or agreement.

Q

QUALITY ASSURANCE

Quality assurance (QA) is an approach to certifying the quality of the output before it is delivered to a customer. It serves as a testament that a project or company exercised due diligence by identifying and removing defects, satisfying mandatory requirements, and assuring the reliability of a product or service. QA can be either qualitative or quantitative in orientation. *Qualitative* is subjective and relies on the application of reliable practices and review of experts. *Quantitative* is objective and relies on statistical analysis and assessment. Ideally, QA should be a function independent from the project itself. It collects data and information on a product and the processes used to build it. To provide an accurate assessment, people involved in QA should have no stake in the projects. Ideally, their assessment should be quantitative in orientation to obtain an accurate account of failures and defect rates and require the occasional unpopular corrective action.

GOALS

- Demonstrate due diligence.
- Ensure that the product or service meets certain requirements and specifications.
- Provide a historical perspective.

OBSTACLES

- Failure to exercise independent judgment
- Desire to take shortcuts in QA audits to expedite the project life cycle
- Failure to distinguish between quantitative and qualitative approaches
- Failure to identify all the standards to test against

STEPS

1. Choose a quantitative or qualitative approach, depending on the project.
2. Conduct tests with an emphasis on maintaining objectivity.
3. Document the results.
4. Develop a corrective action plan, if necessary.
5. Implement the plan.

QUALITY

For a long time, quality meant zero defects, such as in the programs advocated by Philip Crosby. Today, quality is defined as meeting or exceeding customer expectations, which is satisfying the customer. This definition finds its roots in the work of W. E. Deming and J. M. Duran, whose programs such as total quality management (TQM) and total quality control (TQC) have become quite popular. These programs emphasized using facts and data to ascertain customer satisfaction and to encourage employee involvement and addressed cross-functional management. Not surprisingly, quality became associated with tools and techniques for improving customer satisfaction, such as quality circles, metrics, statistical process control, and the PDCA cycle.

Goals

- Minimize defects that result in rework.
- Deliver a product or service that satisfies customer requirements.

Obstacles

- Not establishing a clear definition of customer requirements
- Allowing measurements to become more important than the results achieved
- Using the wrong metrics to ascertain desired results
- Not training people on the tools and techniques necessary to measure and evaluate quality

Steps

1. Develop a definition of quality.
2. Determine the standards for quality.
3. Determine goals and objectives.
4. Develop an overall strategy.
5. Apply the PDCA cycle.
6. Use quality tools and techniques.

R

REENGINEERING

Reengineering is a dramatic overhaul of a business process. The idea is to reengineer a process to obtain a major breakthrough (e.g., 15%) in improvement. It involves discarding the old way of doing business and replacing it with something more efficient and effective. Incremental improvement has little value in the context of reengineering.

GOALS

- Achieve dramatic breakthroughs in results.
- Increase the efficiency and effectiveness of performance.

OBSTACLES

- Trying to fix the existing process under the guise of replacing it
- Not objectively applying criteria to determine the best alternative
- Focusing on technology as opposed to treating it as an enabler

STEPS

1. Document the as-is process.
2. Collect data on the as-is process, documenting the who, what, when, where, why, and how.
3. Review the data to determine non-value-added behaviors (e.g., delays, redundancies).
4. Develop the to-be process.
5. Implement the to-be process, being mindful of resistance to change.
6. Conduct measurements on the to-be process.
7. Make improvements to the new process.

REGRESSION AND CORRELATION ANALYSIS

Regression analysis is used to test hypotheses about the relationships of data within a given population. The basic idea is that, among a given population of data, some data occur independently of other data and some data are dependent on other data. Regression requires identifying the possible relationships between dependent and independent variables. An independent variable precedes the occurrence of a dependent variable, and the inverse condition exists for the dependent variable. Once a hypothesis is formulated, testing begins by collecting data. Differences between what was predicted and what was observed are calculated, confirming or refuting relationships among the data.

GOALS

- Identify the relationships among different variables.
- Draw meaningful conclusions from those relationships.

OBSTACLES

- Assuming that a relationship exists among two or more variables when, in fact, none exists
- Misconstruing the relationship between two variables (e.g., an independent being identified as a dependent variable)
- Assuming that a linear relationship exists between two variables when it is actually nonlinear

STEPS

1. Identify assumptions (e.g., values normally distributed).
2. Identify relationships.
3. Perform testing.
4. Perform calculations.
5. Assess results, confirming or refuting relationships among variables.

REPLANNING

Replanning is revising project plans to meet the goals and objectives of a project. It involves changing the direction of a project. It does not come without a cost in terms of time, money, production, and morale. The biggest danger that accompanies replanning is stopping the momentum of a project to gain a better focus.

GOALS

- Develop a realistic plan.
- Provide confidence that the goal and objectives can be achieved.

OBSTACLES

- Not involving the necessary stakeholders
- Not focusing on the big picture
- Falling into the trap of developing a quick fix
- Not considering hidden costs

STEPS

1. Identify reasons for the replanning effort.
2. Identify exactly what must change.

3. Determine the duration.
4. Identify the stakeholders.
5. Determine the resources and quantities required.
6. Hold individual sessions with stakeholders.
7. Obtain information and suggestions from stakeholders.
8. Hold meetings to resolve differences.

REPORTS

A report is a tool used to display data about a project at a specific point in time. Reports come is all forms and sizes (such as diagrams, charts, and matrices) and contain various levels of content (summary vs. complete documentation).

GOALS

- Provide meaningful feedback on the performance of a project.
- Provide an audit trail for reviews and audits.

OBSTACLES

- Generating too many or too few reports
- Providing too many different versions
- Including invalid or unreliable data
- Lacking ownership
- Lacking consistency

STEPS

1. Determine purpose of a report.
2. Determine the audience.
3. Determine the necessary contents.
4. Determine the schedule of generation.
5. Determine the archiving requirements.

REQUIREMENTS DEFINITION

Requirements definition, which ideally occurs during the early phases of a project, has the ultimate goal of capturing the wants (like to have) and the needs (must have). These requirements are often captured in a document. Among the various views included in requirements definition are the *data view*, which reflects the needs and wants regarding the capture, processing, and delivery of information; the *process view*, which reflects the needs and wants regarding the major processes or functions that occur; and the *behavior view*, which reflects the actions and states of objects. Frequently, all these views are combined into a logical or physical model of the needs and wants of the customer. Ideally, the requirements document should be clear, concise, accurate, and specific. It can be in a graphical or narrative form. It should address technical and business tools, including assumptions, constraints, expectations, priorities, timing, and interfaces. In the end, the requirements document should be signed by all major stakeholders to reflect consensus or agreement and then it should be placed under configuration control.

Goals

- Ensure customer satisfaction.
- Provide a means for assessing product performance later.
- Distinguish between needs and wants.
- Provide adequate scope.
- Enable better product design or service delivery.

Obstacles

- Lack of specificity and completeness when defining requirements
- Failure to document requirements
- Failure to follow requirements
- Failure to put the requirements document under configuration control once agreed upon

Steps

1. Identify the fundamental, key requirements for the project.
2. Collect additional information (e.g., who, what, when, where, why, and how) for the key requirements.
3. Document the requirements.
4. Review the document with the customer for content clarity and accuracy.
5. Determine the methods to capture the requirements.
6. Place the requirements document under configuration control.
7. Use the requirements document as the basis for development activities during the project life cycle.

RESOURCE ALLOCATION

Resource allocation is determining the resources, usually people, necessary to support specific tasks. When allocating resources, it is best to give preference to tasks on the critical path. If tasks are concurrent and on the critical path, give preference to the one with the least total float. For concurrent tasks with the same total float, give preference to the most complex task. The initial allocation of resources can be represented by a set of histograms that profile how the resources are expected to be employed throughout the life cycle of a project. Initial histograms will likely reveal peaks and valleys of an irregular pattern. The goal is to smooth the pattern, known as leveling.

GOALS

- Ascertain how efficiently and effectively resources will be deployed on a project.
- Determine when resources are under- and overallocated.

OBSTACLES

- Not allocating resources efficiently or effectively
- Allocating too many or too few resources
- Competing negatively for resources

STEPS

1. Determine time estimates for tasks at the lowest level of the work breakdown structure.
2. Assign resources to tasks.
3. Print histograms for each resource or categories of resources.
4. Identify peaks and valleys in each histogram.
5. Note points of over- and underallocation.
6. Apply leveling.

RESPONSIBILITY MATRIX

A responsibility matrix is a listing of all people on a project and their respective tasks (Figure 39). It displays the relationship between people and those tasks, as well as the degree of relationship, such as primary or secondary responsibility, when two or more people are involved in a task.

Tasks / Names	Perform A	Perform B	Perform C	Perform D	Perform E
Rogers	P		P	P	S
Ermsloz					
Jesusita	S	S		S	
Wong					
Reichardt		P		S	P
Bergeson					

P = Primary Responsibility

S = Secondary Responsibility

FIGURE 39 Responsibility matrix. (From Project Management Seminar presented by Practical Creative Solutions, Inc., 1996.)

GOALS

- Encourage a sense of responsibility and accountability.
- Clarify roles and responsibilities.

OBSTACLES

- Not indicating level of responsibility
- Providing poor delineation of responsibilities
- Not distributing responsibility matrix to the appropriate people

STEPS

1. Identify team members.
2. Identify tasks.
3. Determine relationships among team members and tasks.
4. In the appropriate cells, indicate the level of responsibility (e.g., primary or secondary), if necessary.
5. Use the work-package-level tasks in the work breakdown structure.
6. Publish the matrix.

REUSE

The concept of reuse, although often associated with software development projects, can play a significant role in many other projects. It entails repeatedly using any pieces or components of something (e.g., a product or its deliverables) to achieve different or similar purposes. Each component must produce consistent and reliable results for each application. Reuse does not just happen. It requires establishing standards (e.g., a common set of rules or model) that address the descriptive characteristics of a component and its respective interfaces, and it requires an overhead responsibility for administering the reuse, such as for tracking components that have been tested and used successfully.

GOALS

- Reduce rework.
- Reduce the tendency to reinvent the wheel.
- Increase delivery speed of a product or service.
- Incorporate quality components.

OBSTACLES

- Not identifying what reusable components exist and scenarios for using them
- Not establishing the administrative infrastructure to apply reuse
- Not establishing or enforcing standards to implement reuse

STEPS

1. Identify the purpose of reuse.
2. Identify candidate components for reuse.
3. Establish an administrative infrastructure (e.g., reuse library) for adequate support of reuse.
4. Establish standards or models for reuse, especially for object structures and interfaces.
5. Maintain version control (e.g., configuration management) of components.

RISK ANALYSIS

Risk analysis is the first function of risk management. It identifies the functions or goals of a project, their relative importance, and the controls that should exist. It lays the groundwork to perform the next function of risk management, risk assessment.

GOALS

- Determine the risks confronting a project.
- Determine the relationships among risks, processes, and goals of a project.

OBSTACLES

- Failure to move beyond identifying risks
- Lack of access to all the necessary data

STEPS

1. Identify the goals and objectives of the project.
2. Identify the major components or elements of the project.
3. Identify existing and "should have" controls for the project.
4. Identify potential risks for the project.
5. Develop a graphic (e.g., matrix) showing the relationships among goals, components, and risks (to include probabilities).

RISK ASSESSMENT

Risk assessment, the second function of risk management, determines the probability of occurrence of a risk and its impact on the overall performance. Specifically, it requires determining the likelihood of occurrence of a risk, or threat, and whether or not existing controls are effective. Often, this determination is made by interviewing and reviewing data.

GOALS

- Determine the adequacy of the relationships among risks, processes, and goals of a project.
- Determine control weaknesses and strengths.

OBSTACLES

- Failing to make an assessment
- Using inconsistent, unreliable data
- Introducing bias into an assessment

STEPS

1. Perform tests to determine the adequacy of controls, using a graphic to show the relationships among goals, controls, components, and risks.
2. Assess the adequacy of controls.

RISK CONTROL

Risk control, the third function of risk management, determines the controls necessary to prevent the occurrence of a risk or to lessen its impact. Controls may be *detective controls*, which disclose risks that surface after occurrence (that is, they reveal that something has happened); *corrective controls*, which help the project recover once a risk has occurred; or *preventive controls*, which stop risks once they surface or mitigate the impact of risks.

GOALS

- Assign responsibilities to implement controls.
- Determine what controls should be added, changed, or removed to augment project performance.

OBSTACLES

- Not identifying the appropriate level of control
- Not assigning responsibility to implement and maintain controls

STEPS

1. Determine improvements for unsatisfactory or nonexistent controls.
2. Identify responsibilities for improvement.
3. Schedule improvements.
4. Establish feedback mechanisms in regard to the effectiveness of the improvements.

RISK MANAGEMENT

Risk management includes the actions pursued by project managers to reduce the impact of a risk to a project. It consists of four functions: risk analysis, risk assessment, risk control, and risk reporting. Risk management addresses *risk*, which is a potential threat to achieving a goal or objective; *control*, which is a measure to prevent or mitigate the impact of a risk if and when it occurs; and *probability*, which is the likelihood that a risk will occur.

GOALS

- Reduce or eliminate budget overruns.
- Offset the negative consequences of budget cuts.
- Reduce the effects of turnover.
- Reduce or eliminate poor quality.
- Reduce or eliminate schedule slides.

OBSTACLES

- Incompleteness
- Failure to identify priorities and probabilities
- Failure to integrate with project outputs (e.g., statement of work, work breakdown structure, and schedule)

STEPS

1. Identify the most important tasks.
2. Document and publish risk management efforts.
3. Encourage involvement of all major stakeholders.
4. Recognize the impossibility of eliminating risk.
5. Conduct risk management on an ongoing basis.

RISK REPORTING

Risk reporting, the fourth function of risk management, involves preparation of a narrative document or presentation on project processes, their risks, their control effectiveness, and recommendations for improvement.

GOALS

- Communicate the results of risk management.
- Provide a record of which controls have proven effective and the ones that require improvement.

OBSTACLES

- Lack of objectivity
- Failure to be complete, clear, or concise
- Failure to follow up on recommendations
- Failure to coordinate review of the document

STEPS

1. Determine the audience for the report.
2. Determine its content.
3. Draft the report.
4. Address positive and negative findings.

5. Provide recommendations for improvement.
6. Review the report for content accuracy, clarity, and conciseness.
7. Publish the report.

S

SAMPLING

Sampling is looking at a subset of an entire population of data and extrapolating conclusions. Several types of sampling are available. *Systematic sampling* is obtaining a sample from a nonstratified population using random selection techniques. *Stratified sampling* is subdividing a population into categories and performing random sampling. *Cluster sampling* is stratifying a sample but sampling from only one stratum. Some key variables to consider in sampling are the size of the sample and population, expected error rate, and the confidence coefficient. Some key concepts of sampling include:

- Every piece of data in a population should have an equal chance of being selected.
- Each piece of data can be selected only once.
- The more variability and less homogeneity in a population, the larger the sample size required.
- More accuracy requires a larger sample size (expected error rate).
- Higher levels of confidence require a larger sample size (confidence coefficient).

GOALS

- Determine a meaningful sample size.
- Determine the best type of sample.
- Provide the basis for probability calculations.
- Draw meaningful conclusions from a sample.
- Reduce the error rate.

OBSTACLES

- Introducing bias through the selection of samples from a population
- Not obtaining a meaningful sample size relative to the size of the overall population
- Not obtaining enough data to perform sampling

STEPS

1. Determine the goals and objectives of a sampling exercise.
2. Determine the population to sample.

3. Calculate the sample size.
4. Collect sample data.
5. Review data and the method for its collection for consistency, reliability, and validity.

SCATTERGRAM

A scattergram reflects the value of a relationship between two variables (Figure 40). A dot on the diagram reflects the occurrence of the relationship. The scattergram consists of an axis representing the potential value of one independent variable and another axis representing the potential value of a dependent variable. A line is drawn through the center of the bulk of the plotted occurrences to reflect the average.

GOALS

- Identify a consistent pattern.
- Identify anomalies.

OBSTACLES

- Not making enough observations
- Assuming a linear relationship when the cause is actually due to an unknown variable

STEPS

1. Determine the independent and the dependent variables.
2. Conduct observations.
3. Draw axes to reflect the relationships between the two variables, and plot the observed occurrence of each variable.
4. Draw a line through the mass of the observations to identify the average.

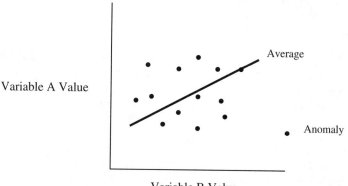

FIGURE 40 Scattergram.

Scheduling

5. Identify any anomalies.
6. Investigate the cause for anomalies.

SCHEDULING

Scheduling involves constructing a diagram that illustrates the expected start and finish of each task in a project and, consequently, the same for the entire project. An additional output from such a schedule is identification of the critical path. Two types of schedules are bar (or Gantt) charts and network diagrams. Bar charts are used for reporting at a higher level, often to a client and senior management. Network diagrams track and monitor performance at a detailed level, often at the project team level. Network diagrams can be either arrow or precedence diagrams. Arrow diagrams, less prevalent today, use descriptions on arrows to describe a task. In addition to the information contained in arrow diagrams, precedence diagrams (used more widely) use boxes or nodes to reflect tasks and other pertinent information. At a minimum, a schedule should provide the following information for each task: early and late start and finish dates, total and free float, a unique identifier (often numeric), duration, and how critical the task is.

GOALS

- Indicate when tasks and an entire project start and finish.
- Show the logical sequence of tasks.
- Identify the critical path.
- Encourage better planning.
- Provide better control.
- Impose discipline throughout the life cycle of a project.
- Create an atmosphere of teamwork.
- Indicate when specific resources are needed.

OBSTACLES

- Lack of scheduling knowledge and expertise
- Lack of time to perform adequate scheduling
- Inadequate resources
- Lack of participation by stakeholders
- Lack of management support
- Fear of commitment

STEPS

1. Develop a work breakdown structure.
2. Perform estimating.
3. Logically tie all tasks together at the work package level with the input of relevant stakeholders.
4. Perform the forward and backward passes using logic, estimates, dependency relationships, and lag values.

5. Review the schedule against requirements described in the statement of work and by the stakeholders.
6. Revise the schedule accordingly.

SCOPE CREEP

Scope creep occurs when a project takes on extra tasks that exceed the original vision (Figure 41). It is one of the biggest killers of projects, particularly high-technology ones. Scope creep can be either explicit or implicit. *Explicit scope creep* occurs when a project deliberately takes on responsibilities or goals that were not originally identified, such as in the statement of work. *Implicit scope creep* occurs when a project inadvertently assumes responsibilities or goals that were not originally agreed to. Often, implicit scope creep occurs because the assumption of a new responsibility or goal is not considered significant at the time but later in the project life cycle the new responsibility becomes more significant. The best medicine for controlling scope creep is to have good change management and configuration management disciplines to evaluate any additional workload.

GOALS

- Reduce the tendency to expand the scope of a project.
- Maintain focus.

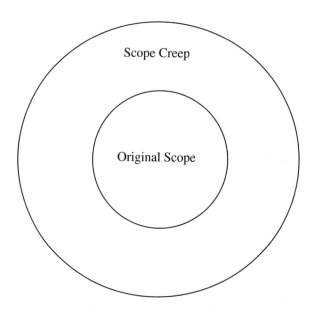

FIGURE 41 Scope creep.

OBSTACLES

- Failure to recognize that the scope has enlarged when dealing with an issue or problem
- The tendency to lose focus on goals and objectives of a project
- An extreme desire to accommodate the client

STEPS

1. Identify the vision, goals, and objectives of the project.
2. Solidify the requirements and specifications.
3. Always evaluate additional requests from the perspective of the vision, goals, and objectives as well as the requirements and specifications.
4. For all requests, except those classified as minor, apply change management and configuration management disciplines.
5. Explain the reasons for rejection.

SECURITY

Security is ensuring the protection of assets from harm and responding when it occurs. It consists of two components: logical and physical security. *Logical security* designs controls from an abstract perspective, regardless of the actual implementation. It is based on the inputs from the risk management activities, contingency planning, and requirements definition. From the logical security comes *physical security*, which is the actual implementation of the chosen logical security. Physical security should address all the major exposures identified earlier and provide the actual controls or measures to be implemented. Some typical exposures that should be addressed by both logical and physical security are errors and omissions, privacy breaches, disasters, physical intrusions of perimeters, and unauthorized access.

GOALS

- Eliminate or reduce the impacts of threats.
- Ensure proper control over resources.

OBSTACLES

- Not integrating security activities with risk management or contingency planning
- Not basing the physical security on the logical security
- Not following up on compliance with security requirements

STEPS

1. Integrate with risk management.
2. Document physical security design.
3. Develop logical/physical security design.

4. Select the best physical design based upon some objective criteria.
5. Implement the physical design.
6. Monitor and take corrective action, if necessary.

SELF-DIRECTED WORK TEAMS

The key concept behind a self-directed work team is empowerment. A self-directed work team operates autonomously, meaning that it requires very little management oversight. It is essentially a cross-functional team of people who determine priorities, allocate resources, select people, and evaluate performance while simultaneously sharing responsibility for results. People on self-directed work teams usually have the requisite levels of skill and knowledge to make decisions that ordinarily require the direct involvement of a manager in other environments.

GOALS

- Empower teams and individuals.
- Encourage a sense of commitment and responsibility.
- Take responsibility for results.
- Eliminate layers of oversight.

OBSTACLES

- Not providing sufficient autonomy for self-directed work teams to succeed
- Not providing the self-directed work team with the skills and knowledge to operate accordingly

STEPS

1. Determine the degree of empowerment to grant.
2. Provide the team with the necessary skills and knowledge.
3. Set boundaries for what decisions and actions are or are not permissible.

SENIOR MANAGEMENT

Senior management is composed of the people responsible for providing guidance and support during management of a project. Specific responsibilities include assigning the project sponsor, allocating resources, prioritizing projects, and providing strategic guidance and direction.

GOALS

- Provide direction to a project team.
- Provide the muscle necessary for a project to survive the politics of project management.

Obstacles

- Not recognizing required roles and responsibilities
- Not providing the necessary support and guidance to help a project succeed
- Failing to isolate a team from political interference

Steps

1. Determine the context of the environment.
2. Determine formal and informal roles and responsibilities.
3. Assess formal and informal powers.
4. Determine role expectations.
5. Identify key members.
6. Determine levels of commitment and support.

SIX HATS

Six-hats thinking, developed by Edward de Bono, is an approach to identifying and applying different modes of thinking to many different activities (e.g., creativity and analysis). *White-hat thinking* is the mode that emphasizes facts, figures, and truth rather than assumptions, emotions, and preferences. *Red-hat thinking* emphasizes emotion or feelings and intuition. *Black-hat thinking* is the negative aspect of thinking, emphasizing the identification of errors and lack of compliance to a standard. *Yellow-hat thinking* is the positive aspect of thinking, emphasizing imagination, speculation, and change. *Green-hat thinking* is creatively developing new ideas and approaches, looking at something from a different perspective, and relying less on structure. *Blue-hat thinking* is disciplined thinking, emphasizing focus, design, and control.

Goals

- Use thinking modes selectively.
- Facilitate thinking about dealing with issues, problems, obstacles, etc.

Obstacles

- Relying on just one or two hat thinking modes
- Applying the wrong mode to a given circumstance

Steps

1. Identify the purpose for which you need to apply six-hats thinking.
2. Ask whether the level of progress is acceptable.
3. If the progress is not acceptable, determine the combination and sequence of hats that will further progress.
4. Remember that the combination and sequence of hats can be changed to suit particular circumstances.

SKILLS MATRIX

The skills matrix is a chart showing the skills required for executing a project and the people who possess them (Figure 42). This information enables a project manager to assign a task to a person who satisfies the skill requirements. Often, a symbol is used to reflect a person's skill level.

GOALS

- Assign a person with the requisite skill to a task.
- Improve productivity.
- Improve morale.

OBSTACLES

- Failure to list all the requisite skills for a project
- Lack of objectivity when assigning a value to reflect level of expertise

STEPS

1. Identify all team members on a project.
2. Identify the desired skills.
3. Collect data about people skills.
4. Build a matrix showing the relationship between the desired skills and the skills that each team member possesses.

Name / Skills	Skill A	Skill B	Skill C	Skill D	Skill E
Smith	1			1	
Jones		2	1		3
Rogers	1	2		2	
Rodero			2		

1 = expert

2 = advanced

3 = basic

FIGURE 42 Skills matrix.

SOCIAL BEHAVIOR TYPOLOGY

The social behavior typology is a tool based on the combination of a person's needs, feelings, thoughts, and actions. Specifically, it requires looking at four types of social behavior manifested as combinations of influencing others (assertiveness) and handling one's emotions (responsiveness). The *analytical style* is demonstrated by a person with low assertiveness and low responsiveness who emphasizes being logical and systematic. The *amiable style* is ascribed to a person with low assertiveness and high responsiveness who emphasizes being cooperative and easygoing. The *driving style* is exhibited by a person with low responsiveness and high assertiveness who emphasizes being determined and decisive. The *expressive style* is exhibited by a person with high assertiveness and high responsiveness who emphasizes being outgoing and spontaneous.

GOALS

- Identify the social types available among stakeholders.
- Apply people in a manner that capitalizes on the category of social behavior.
- Encourage solid teaming arrangements that require the input of social behaviors.

OBSTACLES

- Not understanding the concepts behind the social behavior typology
- Assuming that people generally fall into only one category

STEPS

1. Determine those categories of social behavior that you do or do not exhibit.
2. Observe the behavior of others on your team to determine those categories of social behavior that may be exhibited.
3. Use this knowledge to assign tasks and determine teaming relationships conducive to the project.

SPAN OF CONTROL

Span of control is the number of direct reports that a project manager can handle (Figure 43). In the past, the maximum span of control was 10 people. Historically, the general rule of thumb for span of control has been 7 ± 2 team members; that is, the ideal team size with a leader was anywhere from 5 to 9 people. With the rise in power of computing, the span of control can be expanded to a greater number of people, depending on the circumstances (e.g., tool availability).

GOALS

- Manage people efficiently and effectively.
- Avoid too narrow or broad spans of control.

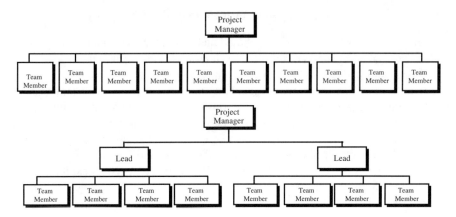

FIGURE 43 Span of control. (From Project Management Seminar presented by Practical Creative Solutions, Inc., 1996.)

OBSTACLES

- Providing too narrow or broad spans of control
- Not gearing the span of control toward the work to be done and the expertise of the people doing the work

STEPS

1. Determine the size of a team.
2. Determine the array of skills required and available.
3. Group team members according to specialty, deliverable, or some other criterion.
4. Determine who will be the team leader based upon knowledge, skill, experience, personality, etc.
5. Ensure that no leader is responsible for more than 10 people.
6. Reflect span of control in an organization chart.

SPEED READING

The volume of documentation on a project seems to expand yearly. From a personal productivity standpoint, about the best way to deal with the volume is to read quickly and effectively. The reality is that only a small amount of documentation and the information contained within it are relevant. The key then is to fish through the documentation and find only what is useful on a project. Doing so requires readers to make arbitrary judgments about what might satisfy their needs and extract what they truly need.

GOALS

- Identify and absorb information quickly.
- Enable more personal efficient personal performance.
- Quickly absorb the main ideas in a document.

Obstacles

- Not distinguishing between essential and nonessential information in reading material
- Not absorbing essential information

Steps

1. Determine exactly what you want to extract from the document (e.g., specific information or general ideas).
2. Conduct a general survey of the document.
3. Note titles and subtitles and diagrams.
4. Skim through the table of contents and index.
5. Skim through the center of each page, noting in pen or pencil, anything that catches your attention.
6. Revisit the sections of interest and read them in more depth.

STAFF MEETINGS

Staff meetings, particularly for large projects, are an effective means to communicate information and share experiences. Unlike other project meetings, such as checkpoint review meetings, staff meetings are held regularly. The session is conducted using an agenda. The agenda typically covers topics that are instructive in nature, such as training overviews and management decisions, as well as the discussion of problems and accompanying suggestions for improvement. At the end, a round robin discussion is held, giving everyone the opportunity to speak. No minutes are taken.

Goals

- Communicate information.
- Build *esprit de corps*.
- Share expertise.

Obstacles

- Failing to follow an agenda
- Allowing meetings to last too long
- Failing to ensure that key people are in attendance
- Allowing certain people to dominate the meeting

Steps

1. Prepare an agenda for the staff meeting.
2. Maintain focus on the agenda.
3. Invite the appropriate stakeholders.
4. Ensure that the proper location is available, replete with supplies, equipment, etc.

5. Ensure that everyone participates during the session, if applicable.
6. Schedule the meetings regularly.

STAKEHOLDERS

A stakeholder is a person or organization having a vested interest in the outcome of a project. Stakeholders may have a direct interest in that they may benefit from the service or product, or they may have an indirect interest, being impacted on only a cursory level. The difficulty is to first identify the stakeholders and then ascertain which of them is directly or indirectly vested in the outcome. The key question to ask is: Does the product or service impact their processes by affecting the productivity or bottom line? If the answer is yes, then these people are stakeholders. Stakeholders in a project environment consist of four generic parties: client (or customer), project manager, project team, and project sponsor. All four may, of course, be considered part of the project team. In this context, the project team is the people who build a product or deliver a service.

GOALS

- Involve the proper participants.
- Encourage ongoing commitment.
- Facilitate communications.

OBSTACLES

- Not identifying all of the relevant stakeholders
- Having the participation of irrelevant stakeholders
- Poorly defining responsibilities and inadequately identifying expectations

STEPS

1. Identify all of the possible stakeholders on a project.
2. Indicate who are primary and secondary stakeholders for senior management, client, team, and sponsor.
3. Delineate stakeholder roles and responsibilities at a high level.
4. Identify stakeholder expectations for involvement and what will result from the project.

STANDARD DEVIATION, VARIANCE, AND RANGE

Standard deviation, variance, and range are calculations to determine the degree of dispersion among data. *Range* reveals the difference between the highest and lowest values in a distribution. It is influenced by extreme values and is used best with small samples and populations of data:

$$\text{Range} = \text{highest value} - \text{lowest value}$$

Standard Deviation, Variance, and Range

For example,

$$56 = 100 - 44$$

Variance is the difference between the observations and the mean, or average, of a sample or population and reflects the average of the squared deviations; the quality of data does not impact the variance:

$$\text{Variance} = \frac{\Sigma(\text{observation} - \text{mean})^2}{\text{total number in sample or population} - 1}$$

- For example, the following table can be used to figure the mean and variance:

	Observation	Observation – Mean	Squared
	11	–6	36
	13	–4	16
	15	–2	4
	16	–1	1
	18	1	1
	22	5	25
	23	6	36
Total	118	–1	119

$$\text{Mean} = 118 \div 7 = 17$$

$$\text{Variance} = 119 \div 6 = 19.8$$

Standard deviation is the square root of the variance. It reveals the average degree of deviation from the mean. Like variance, the higher the calculated value the greater the degree of variation. Ideally, a lower value is sought to reflect less dispersion. Its formula is:.

$$\text{Standard deviation} = \sqrt{\text{Variance}}$$

For example,

$$\text{Standard deviation} = \sqrt{19.8} = 4.5$$

GOALS

- Determine the level of variability with data.
- Draw meaningful, objective conclusions from data.
- Assess the subsequent impact of any changes made.

Obstacles

- Failing to obtain sufficient data observations
- Applying the wrong calculation to a given amount of data

Steps

1. Based upon the amount of available data, determine whether to calculate the standard deviation, variance, or range, or any combination thereof.
2. Draw any meaningful conclusions from the data.
3. Over time, display the results of numerous cycles of calculations in a graph or chart to show trends.

STATEMENT OF WORK

A statement of work (SOW) is a document that defines the scope of a project and the responsibilities of the stakeholders. It provides the basis for developing the plans for a project. The goals and objectives of a project are listed. The scope is defined by noting what is inclusive and exclusive. The deliverables are delineated at high levels as specific products or services to be delivered. The assumptions are documented, including the level of participation. The key stakeholders (e.g., client or sponsor) are listed. The major potential risks and constraints, both business and technical, that a project might face are described. Also included are the project resources involved and to what extent. Significant milestones, or dates, to achieve during the life cycle of a project are listed. The budget is covered, noting the major amounts to spend for each phase or deliverable and for the entire project. Finally, a section is provided for signatures. For projects requiring less formality, a statement of understanding (SOU) is drafted which often does not require as rigorous detail and does not require the advice of legal experts (Figure 44). Often, it is generated for internal projects.

Goals

- Define the goals and objectives.
- Delineate deliverables, responsibilities, etc.
- Identify and resolve points of disagreement and ambiguity early.
- Have the SOW serve as an agreement among stakeholders.

Obstacles

- Caving in to an unrealistic request made by a stakeholder
- Including unclear phrases and terminology
- Filing the SOW away and never reviewing it
- Failing to delineate all items of major importance
- Failing to obtain stakeholder agreement

I. INTRODUCTION
This project resulted from the failure of our sales office to meet the changing needs of our clients. Relations between our company and existing clients have become strained, creating a 30 percent drop in our sales volume. The Board of Directors has requested that a team of top salespersons and marketing professionals develop strategies and tactics for reversing this downward trend in sales.

II. SCOPE
This project will not involve implementation of any of the strategies and techniques resulting from its efforts. The major goals and objectives of the project are to:
- Develop a better understanding of the clients' needs
- Identify ways to improve relations
- Identify why competitors are doing much better than we
- Produce a detailed document (proposal) listing and discussing strategies and techniques, including ideas on how to implement them

The project will staring on April 19, 20XX and finish on August 21, 20XX. The total cost must not exceed $150,000.

III. RESPONSIBILITIES
    A. Sales and Marketing Director will:
        1. Appoint a five-member team to work on the project
        2. Serve as the project manager
        3. Prepare a final report
        4. Report weekly to the corporate president on the status of the project
    B. The President will:
        1. Review the proposal for completeness
        2. If approved, give a presentation on the results to the Board of Directors
        3. Upon approval of the Board of Directors, develop plans for implementing the recommendations

IV. AMENDMENTS
This document may change according to change control procedures established in the corporation.

V. APPROVALS

_____   _____
Project Manager   President

_____   _____
Date   Date

FIGURE 44 Statement of understanding. (From Project Management Seminar presented by Practical Creative Solutions, Inc., 1996.)

STEPS

1. Draft the SOW.
2. Submit copies for stakeholder review.
3. Hold a joint meeting with stakeholders to discuss discrepancies and resolve differences.
4. Revise the SOW.
5. Hold additional review sessions with stakeholders, if necessary.
6. When the SOW is acceptable, date and sign it.

STATISTICAL PROCESS CONTROL

Statistical process control (SPC) is a graphical technique that displays observational data about the performance of a product or service over time (Figure 45). The idea is to capture variation and determine its significance. Once identified, the source of variation is also identified and fixed. SPC requires an upper control limit (*ucl*) and lower control unit (*lcl*). Determination of the *ucl* and *lcl* depends on the statistical variations at ±3 standard deviations. A line is drawn to reflect the average, or mean. A variation occurs when any observation falls outside of the *ucl* or *lcl*. The data are nonconforming when seven successive observations go beyond the *ucl* and *lcl*.

GOALS

- Identify variation.
- Diagnose the cause for any variation.

OBSTACLES

- Not enough observations
- Failure to identify and take action on nonrandom variation
- Treatment of random variation as nonrandom variation and vice versa

FIGURE 45 Statistical process control.

STEPS

1. Determine the process to observe.
2. Identify the characteristics to measure.
3. Determine the expected average for observations taken over time and draw a line.
4. Identify the upper control limit.
5. Identify the lower control limit.
6. Take measurements.
7. For observations that consistently go beyond *ucl* or the *lcl*, conduct an analysis regarding the cause.
8. Take corrective action, if necessary.
9. Take measurements to determine the effectiveness of any corrective action.

STATISTICS

Statistics is collecting, analyzing, and evaluating data. Data can be discrete or continuous variables. Discrete variable data have distinct differences among values. Continuous variable data can fall anywhere within a spectrum. All instances of variables should have a value with each occurrence, a frequency of occurrence, and a probability of occurrence. Each piece of data can have a dependent or independent relationship that is linear, multilinear, or nonlinear. Data can be measured in several ways, using different scales. Nominal scales use names or symbols that do not lend themselves to mathematical manipulation. Ordinal scales are rank ordering according to some type of magnitude. Interval scales establish meaningful differences among categories of data. Ratio scales use a set of numbers to represent a point on a scale. A measurement of data should satisfy three criteria: A measure should be reliable; that is, it should provide no variation when measuring the same phenomenon each time. A measure should be applied consistently; that is, it should be administered without any variability. A measure should be valid; that is, it should measure what it purports to measure.

GOALS

- Obtain objective data.
- Use data to ascertain and assess performance.
- Objectively determine the impact of any actions.

OBSTACLES

- Introducing bias into the data collection or calculations, or both
- Not considering the threats to data validity
- Using the results of statistical calculations to mislead
- Not collecting data using reliable, consistent methods
- Misinterpreting a dependent variable as an independent variable and vice versa

STEPS

1. Determine the objectives behind the statistics to calculate.
2. Determine the method of data collection.
3. Select the appropriate tool or technique.
4. Collect the data.
5. Seek reliability, consistency, validity, and objectivity during data collection and calculations.
6. Display the results in a manner that provides an objective portrayal.

STATUS ASSESSMENT

Status assessment is how well a project progresses up to a given point in time for schedule, budget, or quality criteria. The idea is to determine whether a project has progressed as planned. If not, then some decision must be made on how to ensure that plans and reality match each other as closely as possible.

GOALS

- Convert data into information.
- Determine the level of success for a project.

OBSTACLES

- Not using a reliable, repeatable approach
- Reluctance to make an assessment
- Using dated, incomplete information
- Entering bias into an assessment

STEPS

1. Obtain data from the status collection effort.
2. Generate reports reflecting desired metrics.
3. Review information to determine the degree of variance, if it exists.
4. Determine whether corrective action is necessary.

STATUS COLLECTION

Status collection involves the use of tools and techniques to determine how well certain tasks and an entire project progress according to schedule, budget, and quality criteria. Often the most common tools and techniques are meetings (such as checkpoint review and status review meeting) and forms. To be effective, status collection requires performing in an atmosphere of trust, which requires a willingness by all stakeholders, especially the project manager, to receive a status report that tells them what they need to hear, not want they want to hear.

Goals

- Obtain the necessary data to generate meaningful information for reports.
- Provide a consistent, reliable means to identify and assess progress throughout the life cycle of a project.

Obstacles

- Not performing consistently
- Not using a reliable approach
- Not providing adequate coverage

Steps

1. Determine data requirements.
2. Determine the means (e.g., forms) to collect data.
3. Determine the schedule for collecting the data.
4. Determine who will provide the data.

STATUS REVIEW MEETING

The purpose of a status review meeting is to examine the outcome of status collection on cost, schedule, and quality. It is held after completing status collection. It is not held during the meeting. What status is given is incidental during the assessment of current progress. An agenda is followed and minutes are taken. When assessments are negative, attendees determine corrective actions and workarounds to execute to reverse the trend. Status review meetings are best held regularly, such as weekly or biweekly. Minutes are taken and distributed after each session.

Goals

- Assess progress of individual tasks and an entire project.
- Determine corrective actions or workarounds.
- Keep a project focused on the goals and objectives.
- Share information.

Obstacles

- Failing to follow an agenda
- Not taking minutes
- Failing to collect status before the meeting
- Not having all the appropriate stakeholders present
- Allowing certain people to dominate the meeting

Steps

1. Prepare an agenda.
2. Maintain focus on it.

3. Invite the appropriate stakeholders.
4. Collect status before the meeting.
5. Ensure that the proper location is available, replete with supplies, equipment, etc.
6. Ensure that everyone participates during the session, if applicable.
7. Concentrate on facts and data.
8. Determine corrective actions and workarounds, if necessary.
9. Take and publish minutes.
10. Schedule regularly.

STEWARDSHIP

Popularized by Peter Block, stewardship for organizations in general and projects in particular emphasizes governance over control. That means placing less emphasis on control, paternalism, and patriarchy to promote consistency and predictability. Instead, stewardship requires empowering people less through control and more through persuasion and garnering commitment via a sense of personal responsibility to achieve results. This perspective moves away from the idea of the heroic project manager who is at the center of everything to becoming one of a steward, who harnesses the potential of an organization (in this case, a project) to achieve results.

GOALS

- Build commitment.
- Encourage a sense of responsibility.
- Increase trust and ownership.
- Foster a sense of partnership.

OBSTACLES

- Confusing stewardship with easygoing management
- Failing to "walk the talk" when espousing principles behind stewardship
- Falling back on controlling when a project fails to progress exactly as planned

STEPS

1. Determine the existence of paternalism on a project.
2. Ascertain the practices that are "overcontrolling" (e.g., too many constraints) project execution.
3. Identify opportunities for encouraging stewardship by removing constraints (e.g., overly restrictive procedures, excessive oversight by management).
4. Encourage joint accountability for deliverables and results.
5. Provide an environment for communicating and sharing information.
6. Encourage people to take the initiative.
7. Inculcate a sense of partnership with all stakeholders.

STRATEGIC PLANNING

Strategic planning is determining the vision of a company and the corresponding strategy to achieve it. It requires looking at strengths, weaknesses, trends, and risks. It also requires determining priorities, allocating resources, and exploring future business opportunities. The result is a strategic plan that directs subsequent tactical actions and operations throughout a time period, such as a year. The result is a comprehensive strategic plan for an entire company. Projects play an important role in the implementation of a strategic plan both at the strategic and tactical levels. It is important, therefore, that the goals and objectives of a project are aligned with those of the company; otherwise, the project will likely not provide value to the bottom line. Projects, whether entrepreneurial or administrative in nature, should directly link to the vision embodied in a strategic plan. It is especially imperative that they do so at the executive or cross-functional level of a corporation. All project plans, therefore, should reference applicable portions of a strategic plan.

GOALS

- Promote a relationship between strategic planning and project management.
- Ensure that projects achieve overall strategic goals and objectives of a firm.

OBSTACLES

- Lack of a connection between a strategic plan and project plans
- Failure to determine how projects contribute toward achieving the goals and objectives of a company

STEPS

1. Develop or refer to the strategic plan.
2. Map the goals and objectives of the strategic plan and the project.
3. Map any relevant additional information between the strategic plan and project, such as:
   a. Threats
   b. Opportunities
   c. Priorities
   d. Trends
4. If a gap exists between the strategic plan and the project plan, consider whether to remove anything.

SUPPLIER MANAGEMENT

Supplier management is a buyer managing his or her relationships with suppliers. The relationship should strive to be mutually beneficial through a sense of fair dealings and of partnership. It should not be mistrustful and adversarial. The ingredients for good supplier management are having a good assessment of a supplier's capabilities; conducting a survey of multiple suppliers; coordinating with other departments, such as purchasing and legal, in a buyer's firm; maintaining ongoing communications with suppliers; monitoring relationships; and taking corrective action for substandard delivery of products or services.

Goals

- Obtain the most reliable, qualified suppliers.
- Develop a meaningful partnership with suppliers.
- Manage the relationship with suppliers.

Obstacles

- Treating suppliers as short-term "add-ons"
- Not following any quality standards when dealing with suppliers
- Failing to conduct a meaningful survey of suppliers
- Failing to maintain ongoing communications

Steps

1. Determine the products or services to be provided by suppliers.
2. Develop standard parts or products lists, if applicable.
3. Determine the desired capabilities of particular suppliers.
4. Evaluate the overall relationship with specific suppliers.
5. Develop an approved suppliers list, if possible.
6. Coordinate with other internal departments, such as purchasing, contracts, and legal.
7. Negotiate a contract with clearly defined terms and conditions.
8. Maintain ongoing communications with suppliers.
9. Collect metrics on the performance of suppliers.
10. Monitor relationships with suppliers, taking corrective action if necessary.

SUPPLY CHAIN MANAGEMENT

The concept behind supply chain management is taking a holistic, integrated view of the major processes and systems in a firm from concept to customer support. It requires looking at, coordinating, and integrating a host of functions, such as materiel, production, logistics, and distribution. Of special interest are the roles and responsibilities of major players, product and information flow, inventory levels, sales trends, forecasting, and production planning, as well as considering requirements and constraints. The ultimate goal of supply chain management is to satisfy customers by providing the right product at the right time at the right place and in the right condition. A supply chain perspective plays a key role in the implementation of enterprise resource planning (ERP) and e-commerce.

Systems Theory

Goals

- Ensure customer satisfaction.
- Perform more efficiently and effectively.
- Recognize the impact of actions throughout the supply chain.

Obstacles

- Failure to identify all the major stakeholders and processes in a supply chain
- Failure to identify all the important variables in a supply chain (e.g., customer requirements)

Steps

1. Take a systems perspective of the supply chain.
2. Identify the major processes occurring in the supply chain.
3. Identify the major constraints affecting those processes.
4. Identify major systems.
5. Identify major stakeholders.
6. Conduct an impact analysis of any changes on the supply chain.
7. Account for information flow.

SYSTEMS THEORY

Systems theory is concerned with building a model of how something (e.g., an entity or idea) functions. Generally, it is a broad, general description because it cannot capture everything. Any systems theory will have these elements: input, outputs, components, feedback mechanisms, open or closed boundaries, participants, rules (including constraints), patterns of behavior, and anomalous behavior. It is generally perceived that all systems are goal oriented and require integration among all the components via some type of medium. Two key concepts are that a system maintains equilibrium (homeostasis) and that over time the system may degrade in performance (entropy).

Goals

- Identify and map the elements and relationships of a system.
- Predict the behavior of an actual system under given circumstances.

Obstacles

- Construing the behavior of the system being modeled to be law
- Inadequately identifying elements and their relationships
- Failing to accept the reality that a model can be no more than an approximation

STEPS

1. Define:
   a. Components
   b. Relationships among components
   c. Medium (e.g., data, signals) for executing relationships
   d. Major players (e.g., people, organizations)
   e. Boundaries (e.g., open or closed)
   f. Business rules
   g. Critical relationships
2. Determine the different scenarios for executing the rules.
3. Identify:
   a. Constraints (e.g., policies, resource limitations)
   b. All patterns of behavior
   c. Communication mechanisms for feedback

T

TEAM BUILDING

Team building is acting to ensure that stakeholders collaborate efficiently and effectively to achieve the goals and objectives of a project. It requires getting people to focus on the goals and objectives, share information and other resources, act in the best interests of the project, and recognize the impact of their actions on others.

GOALS

- Reduce negative conflict.
- Turn negative conflict into a positive experience.
- Encourage a win–win result for all stakeholders.

OBSTACLES

- Conflict over resources
- Allowing negative politics to take over
- Failing to communicate effectively
- Failing to engender *esprit de corps*

STEPS

1. Institute effective spans of control.
2. Appoint leadership, if necessary.
3. Assign responsibilities.
4. Maintain accountability.
5. Encourage camaraderie.
6. Provide effective communications.
7. Institute unit of command.
8. Provide for a good work environment.

TEAM ORGANIZATION

To maximize the productivity of their teams, project managers must establish a managerial structure with one of two structures: a matrix or a task force structure, with well-defined reporting and delineated roles and responsibilities.

GOALS

- Employ people efficiently and effectively.
- Encourage a sense of responsibility and accountability.
- Clarify responsibilities and accountabilities.

OBSTACLES

- Violation of the unity-of-command principle
- Over- or underextending span of control
- Not clarifying roles and responsibilities
- Inappropriate assignment of roles and responsibilities according to skills, knowledge, experience, etc.

STEPS

1. Pick the most appropriate team structure (e.g., task or matrix) and adapt it to circumstances.
2. Apply the unity-of-command principle.
3. Apply effective span of control.
4. Publish related documents (e.g., organization chart).

TEAMING BASICS

The difference between a team and a group or committee is that a team is focused on achieving specific results over a short time period. Ideally, an effective team is what Jon R. Katzenbach and Douglas Smith refer to as high performance teams. There are various ways to achieve high performance teams but most have these items in common: a vision and plan emphasizing performance objectives; defined but flexible roles and responsibilities; information sharing through ongoing communications; an ever-present sense of ownership, commitment, and mutual accountability; a disciplined, common approach to achieve desired results; a high level of trust on the team and with management; and an overall tolerance for ideas and mistakes. Above all, an overriding sense of synergy exists as people shift from thinking in terms of "my" work to "our" project.

GOALS

- Achieve results efficiently and effectively.
- Focus on achievements.
- Increase synergy.
- Build *esprit de corps*.

OBSTACLES

- Concentrating on one aspect of a team (e.g., responsibilities only)
- Providing an unclear, uncommitted vision

- Failing to establish a common approach for achieving results
- Ignoring intangible factors that contribute to a team's success

STEPS

1. Identify a common vision for a project.
2. Define roles and responsibilities.
3. Encourage a sense of commitment and sponsorship.
4. Encourage communications and sharing of information.
5. Build mutual accountability.
6. Offset any tendency towards groupthink.
7. Provide a disciplined, focused approach toward achieving the vision.

TECHNOLOGY TRANSFER

Technology is changing and doing so rapidly; however, the changes are not only occurring with the technology itself. New technology impacts other aspects of projects, such as people, processes, performance, and, ultimately, profits. Technology transfer focuses on implementing automated tools in the work environment. It requires careful planning and flexible implementation so that positive impacts exceed negative ones. Technology transfer also requires getting people to change because they want to and not have to. The idea is to identify the internal and external impacts to ensure that the culture and other key environmental circumstances are not overlooked. Of primary importance is determining the change management roles and their accompanying responsibilities as well as getting trust and commitment.

GOALS

- Encourage people to embrace new technologies.
- Minimize the negative impacts of technology transfer.
- Identify the accompanying risks and opportunities.

OBSTACLES

- Failure to identify key impacts of new technologies
- Inability to capitalize on the accompanying opportunities
- Failure to plan appropriately
- Failure to appreciate the people side of new technology
- Failure to consider feedback

STEPS

1. Recognize that technology transfer affects people and processes.
2. Conduct an impact analysis (including risks) of the technology transfer.
3. Develop an overall strategy to achieve goals and objectives.
4. Develop a plan to achieve goals and objectives and execute the strategy.
5. Implement the plan, keeping in mind the need for feedback.

6. Identify major stakeholder roles involved in the technology transfer (e.g., change agent, advocate, champion).
   7. Identify constraints.
   8. Encourage ownership and commitment.

TELECOMMUTING

The dramatic advancements related to the Internet and microcomputers have provided people with the opportunity to work from home, known as telecommuting. Under the proper circumstances, telecommuting can serve as powerful means to increase both productivity and morale. By having the necessary tools and level of latitude, telecommuters can focus on their work by avoiding the interruptions and management intrusion that often occur in an office. It is important to establish parameters up front for what is acceptable and unacceptable when telecommuting.

GOALS

- Engender a sense of responsibility in the telecommuter.
- Provide flexible work hours.
- Minimize interruptions.

OBSTACLES

- Possibly higher upfront costs
- Reluctance by management to trust telecommuters
- Telecommuters working harder for longer hours due to a greater sense of responsibility

STEPS

   1. Determine who can telecommute.
   2. Define processes and procedures to follow when telecommunicating.
   3. Provide the necessary resources for people to telecommute.
   4. Ensure that opportunities exist to bring team members together to meet periodically in person.
   5. Set aside time to meet with each telecommuter to ensure ongoing communications.
   6. Provide the autonomy for people to telecommute effectively.

TESTING

Testing involves identifying flaws in a product or service. Often, testing is performed throughout various phases of a project but more often toward the end of its life cycle. The ultimate goal of testing is to find errors, not prove correctness. It involves developing an overall testing strategy, testing scenarios or cases, testing approaches or techniques, and testing completion criteria. Testing is not just on the product or service itself but also the processes that produce them. Essential to executing an effective test plan are having testers who are independent from the act of producing the product or service and can objectively assess the results. Recently, the emphasis has been on reducing testing by incorporating quality principles in the actual process of building a product; however, this does not negate the need for testing but rather changes the degree and type of involvement for testing.

GOALS

- Identify defects or errors early.
- Reduce the cost of the product life cycle.
- Prevent future legal complications.
- Determine the source of any defects or errors.

OBSTACLES

- Shortcutting testing to expedite the project life cycle
- Testing for success rather than failure
- Not establishing criteria to use for testing
- Not identifying all the test cases and scenarios
- Failing to follow up on test results

STEPS

1. Develop an overall test strategy.
2. Determine test cases and scenarios.
3. Determine test criteria.
4. Test for errors and failure.
5. Document test results.
6. Develop a corrective action plan.
7. Implement the plan.
8. Maintain objectivity in testing through segregation of duties and use of test criteria.

TIME MANAGEMENT

Time is often a premium for projects. Many internal and external factors pull for the attention of all stakeholders and particularly for project managers. Just about all of the demands require immediate attention, especially from the perspective of an originator; however, the laws of physics make that impossible. It is imperative that project managers set priorities and then determine what receives attention first, second, and so on. The best way to set priorities is to evaluate a request from the perspective of the goals and objectives of a project. This perspective will maintain focus and allow pro-action rather than reaction to demands. Above all, project managers must say "no" with discretion and perform screening.

GOALS

- Use time efficiently and effectively.
- Maintain focus.
- Determine what is realistic to do under given circumstances.

OBSTACLES

- Inability to set priorities
- Become too focused on satisfying people's demands
- Inability to say "no"

STEPS

1. Determine the criteria to establish priorities.
2. Apply priorities.
3. Set time aside to address major priorities.
4. Recognize what you can or cannot do or control.
5. Be willing to say "no" discreetly to certain requests, if necessary.

TOP-DOWN AND BOTTOM-UP THINKING

Top-down and bottom-up are two opposite approaches taken to design or build just about anything (Figure 46). Top-down thinking requires taking a holistic perspective and then exploding it into components from general to specific. Often, it requires going through several iterations until the explosion gets down to an acceptable level of detail, e.g., moving from general to detail. Top-down thinking is often used to design and develop work breakdown structures, architectures, etc. Bottom-up thinking requires taking a specific, piece-by-piece approach and gradually working one's

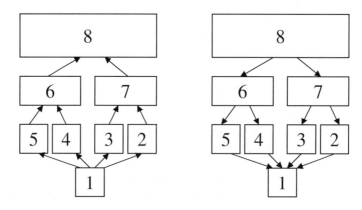

FIGURE 46 Top-down and bottom-up thinking. (From Project Management Seminar presented by Practical Creative Solutions, Inc., 1996.)

way up to higher levels (e.g., moving from specific to general). Bottom-up thinking is often best used when the overall picture is lacking and must be incrementally determined by adding, removing, or altering something, such as during prototyping.

GOALS

- Design or develop a cohesive product.
- Design or deliver a complete service.
- Deal effectively with too little or too much data and information.

OBSTACLES

- Failing to understand when to take which approach
- Using the wrong approach in a given context
- Not knowing when a certain approach is thorough enough

STEPS

1. For top-down thinking:
   a. Start at a high level
   b. Determine the best way to chunk it
   c. Explode each chunk down to a desired level of granularity
2. For bottom-up thinking:
   a. Start at a very low level of granularity
   b. Progress up to a higher level of abstractness
   c. Continue progressing up to a higher level of abstractness until comfortable with a level that conceptually encapsulates everything under it

TOTAL FLOAT

Total float is the amount of time that a task can slide before it impacts the project completion date of a project. It is calculated as the difference between the early finish date and the late finish date. The path in a network diagram that contains the least amount of float (ideally, having a value of 0) is known as the critical path; hence, the basic rule is that the lower the float, the more critical the task from a schedule perspective.

GOALS

- Determine the critical path.
- Facilitate the allocation of resources.

OBSTACLES

- Not recognizing the relationship of float to the critical path
- Not recognizing the difference between free and total float
- Not recognizing that float changes constantly

STEPS

1. Obtain a network diagram.
2. Identify the early and late start and finish dates.
3. For each task, subtract the early finish date from the late finish date (LF – EF) to derive the total float.
4. Note that the tasks with minimum float indicate critical path and, consequently, cannot slide.

TRACKING AND MONITORING

Tracking is reviewing past performance according to schedule, budget, and quality criteria. It involves looking at what has been done previously up to a given point in time. In other words, tracking looks at the past. Monitoring looks into the future and projects future results according to schedule, budget, and quality criteria. It involves using past performance up to a given point in time and projecting how the current project will end up, assuming past performance and considering the remaining work to be done.

GOALS

- For tracking, understand where the project has been and where it is.
- For monitoring, understand existing conditions to anticipate and when a project will end (e.g., ahead or behind schedule).

OBSTACLES

- Using dated, incomplete data
- Using an inconsistent, unreliable approach
- Providing inadequate coverage

STEPS

1. For tracking:
   a. Collect status regularly
   b. Generate appropriate reports
   c. Evaluate contents of reports
   d. Determine if a significant variance exists between current and planned situations
2. For monitoring:
   a. Take the results from tracking
   b. Determine the degree of variance from projected performance
   c. Determine any corrective action to take to close the variance between planned, actual, and projected

TRAINING

Training is useful for reducing the effects of a poor learning curve on a project. To obtain maximum payback from training, it is important to implement training in a cost-effective manner. Time is often tight on a project and time spent on training should be viewed as an investment. Training should be planned and executed with specific goals and objectives in mind. The key is to conduct a thorough needs assessment to ensure that training is timely and relevant. The assessment should address topics such as the subject matter, equipment, facilities, personnel, materials, goals, objectives, types of training (e.g., online, workshop), and learning styles. Too often, training is conducted on a project without any clear requirements, which only leads to poor payback. A thorough needs assessment goes a long way toward preventing this from happening. Frequently, too, few trainees apply their new knowledge or skills on a project, either because of fear or because the opportune time has come and gone. It is important, therefore, to ensure that people receive training in advance to be able to apply it to a project.

GOALS

- Reduce learning curve.
- Provide skills to perform more effectively and efficiently.
- Reduce the impact of turnover and downsizing.

OBSTACLES

- Failing to be patient with the learning curves of other people
- Not recognizing the need to apply new learning quickly to reinforce concepts, principles, and techniques

STEPS

1. Conduct a training needs assessment.
2. Determine the different learning styles to accommodate.
3. Develop a training plan.
4. Prepare for training.
5. Conduct the training, focusing on its goals and objectives.
6. Provide opportunities as soon as possible to apply what was learned during training.

TRANSACTIONAL ANALYSIS

Transactional analysis (TA) is a psychological approach for understanding the ways people interact with each other. The key is to recognize that three states are identified that represent patterns reflective of attitudes, feelings, and experiences. The *parent ego state* represents critical, controlling parents. The *adult ego state* represents intelligent, reasonable people who weigh circumstances. The *child ego state* represents the natural impulses of a child often exhibited immaturely. The various combinations of the ego states result in transactions, such as when one person acts in a parent ego state and another in a child ego state. As an example, parent–child transactions that have predictable results are called complementary, while parent–child transactions that are unpredictable are called cross-transactions.

GOALS

- Identify the type of transactions occurring on a project among stakeholders.
- Apply people in a manner that reduces negative conflict.
- Encourage solid teaming arrangements.

OBSTACLES

- Not understanding and applying all the principles behind TA
- Assuming that people fall into only one ego state

STEPS

1. Observe the transactions of people on your team from the perspective of the three states.
2. Use this knowledge to assign tasks and to determine teaming relationships that are conducive to the scripts that people use.
3. Determine the positive and negative impacts of your behavior, particularly with respect to whether you transact from a parent, adult, or child perspective.

TUCHMAN MODEL

Developed by Bruce Tuchman, this model identifies four phases through which a team evolves, each phase preceding the next. The *forming phase* occurs when a team initially forms as a disparate association of people. The *storming phase* occurs when team members start working things out by discussing and defining details to get started. The *norming phase* occurs when a team is ready to achieve its goals through cooperation and with a sense of organization. The *performing phase* occurs when a team is totally focused on accomplishing a goal while simultaneously being receptive to new ideas and approaches.

GOALS

- Recognize that teams go through a life cycle.
- Marshal the energies of a team at a particular point in time.
- Be patient with teams as a means for achieving goals and objectives.
- Apply people's talents during specific stages.

Tuchman Model

Obstacles

- Not recognizing the key applications of the stages of a team
- Not managing the progression of a team through each of the four stages

Steps

1. Learn the characteristics of the four phases.
2. Recognize the positive and negative characteristics of each phase.
3. Determine the phase of the project.
4. Leverage the positive aspects of each phase.
5. Take measures to offset negative aspects of each phase.
6. Determine ways to progress from one phase to the next.
7. Understand that some projects progress through some phases faster than others.

U

UNITY-OF-COMMAND PRINCIPLE

The unity-of-command principle states that a person can report to only one boss and cannot, so to speak, serve two masters. Following the unity-of-command principle can be extremely difficult on a project that exists in cross-functional, matrix environments. One person might work for a project manager for work requirements while simultaneously reporting to a functional manager for pay, promotion, etc. Obviously, such violation of the unity-of-command principle can lead to confusion and confrontation because directions can be at cross purposes. The best approach for overcoming this scenario is for the project manager and the functional manager to communicate with each other to reduce the impact on any team members.

GOALS

- Use team members efficiently.
- Reduce frustration and negative conflict.

OBSTACLES

- Unwittingly violating the unity-of-command principle and doing nothing about it
- Failing to acknowledge and address the negative impacts when the unity-of-command principle is violated

STEPS

1. Minimize any circumstances for encouraging the violation of the unity of command principle.
2. If a violation is necessary, establish ways to offset the negative impacts to the individual and the project.

V

VARIANCE

Variance is the difference between planned and actual performance, or planned performance minus actual performance. Two of the most common variances occur with schedule and cost. *Schedule variance* is the difference between planned start and end dates and actual start and end dates for each task and the entire project. The typical formula is planned start/end date minus actual start/end date. Of course, a negative value indicates a possible problem. *Cost variance* is the difference between budgeted costs and actual costs incurred for each task and its entire project, or budgeted cost minus actual cost. A negative variance indicates a cost overrun.

GOALS

- Determine the difference between what was planned and what has occurred.
- Identify ways to reduce variance.
- Determine the source of negative and positive variance.

OBSTACLES

- Forgetting that a variance indicates an anomaly, not necessarily something wrong
- Not recognizing the need to take corrective action regarding schedule, budget, or quality

STEPS

1. Determine the baseline of cost, schedule, and quality.
2. Review reports from a status assessment, looking for deviations from baselines (variance) in respect to cost, schedule, and quality criteria.

VIDEOCONFERENCING

With today's powerful information technology readily available, videoconferencing has become a more common phenomenon. It allows for better communication among team members who are spread over great distances; results in less traveling; and allows team members to become familiar with others involved on the project. Videoconferencing can initially be expensive, especially when establishing videoconferencing rooms located over a wide geographic area. Fortunately, thanks to microcomputers that feature greater power and more sophisticated software, videoconferencing can be conducted less expensively, although degradation in performance can occur. Videoconferencing makes sense when travel is too expensive, when personal interaction is not required, when security of data and information is not a priority (unless sophisticated encryption is used), and when specific questions must be answered. It does not work very well when cultural issues are involved, when a personal dialog of an intimate or secret nature is necessary, or when the costs of travel are less expensive (particularly in regard to using large-scale videoconferencing rooms).

GOALS

- Provide greater communication.
- Enable greater collaboration.
- Reduce travel costs.

OBSTACLES

- Using technology that is not mature, resulting in signal delay
- Using teleconferencing as a means for displacing human interaction

STEPS

1. Provide the resources for videoconferencing.
2. Recognize the shortcomings that accompany videoconferencing.
3. Identify when videoconferencing is more effective than personal interaction.
4. Ensure that the technology is mature enough to deal with the topic being discussed.

VIRTUAL TEAMING

With the current advances in Internet technology, team members can work farther apart and still make a significant contribution to the outcome of a project. Someone in Rome and another person in New Delhi can work together to complete a project without ever actually meeting. Activities can occur during all hours of the day. Virtual teaming requires considerable collaboration, coordination, and communication as a project progresses throughout its life cycle. When virtual teaming is successful, the biggest advantage is the tremendous reduction of overhead expenses related to facilities and travel. If it fails, then the costs can be prohibitive due to rework, delays, and schedule slides.

GOALS

- Encourage collaboration over greater than usual physical distances.
- Encourage better communication for team members not working together in the same physical location.

- Reduce overhead expenses.
- Engender a greater sense of trust by stakeholders and assumption of responsibility by the people doing the work.

Obstacles

- Treatment of people not present as "out of sight, out of mind"
- Lack of compatibility between hardware and software
- Poor quality of work due to improper oversight
- Lack of synchronized timing of deliverables with distant team members

Steps

1. Identify all members of the virtual team.
2. Encourage ongoing communications among all members.
3. Provide opportunities for people to get together physically.
4. Hold people responsible for discrete deliverables.
5. Ensure that all team members have compatible equipment.
6. Conduct periodic one-on-one sessions with each person.

WINDING DOWN

Winding down is an important activity of the controlling function. It requires acting with resources, especially people, to minimize interruption and sustain productivity. This activity is especially important for projects with a task force structure. Team members may find themselves without opportunities for work after completing a project.

GOALS

- Maintain the momentum of the project toward its goals and objectives.
- Maintain stakeholder commitment to the very end of the project life cycle.

OBSTACLES

- Loss of focus on goals and objectives
- Lack of commitment by certain stakeholders
- Lack of follow-through by certain stakeholders

STEPS

1. Release or reassign people as early as possible.
2. Obtain formal approval of the final product or service from the client.
3. Recognize stakeholders who have performed above the norm.

WORK BREAKDOWN STRUCTURE

A work breakdown structure (WBS) is a detailed listing of the tasks required to complete a project (Figures 47 to 50). It serves as the basis for estimating, creating schedules, and assigning responsibilities. Often, it is approached in one of three ways: by responsibility, by deliverables, or by phase. Regardless of approach, a WBS has certain characteristics. The contents have a top-down, general-to-specific flow. Each item has a unique numeric designation that indicates its position within the WBS. The items in the higher level portion of a WBS are described using an adjective and a noun. The items in the lower level portion of a WBS are described using a command, or action verb, plus an object. Ideally, each leg in a WBS is exploded into finer detail until a specific task can be done in 80 or less hours. This is known as the 80-hour rule. Generally, two sources exist for developing a WBS. The first is

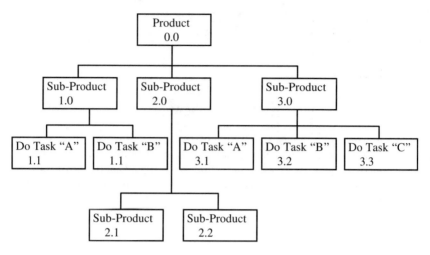

FIGURE 47 Typical work breakdown structure. (From Project Management Seminar presented by Practical Creative Solutions, Inc., 1996.)

existing documentation (such as statements of work, studies, reports, organization charts, policy statements, procedures, work breakdown structures from similar projects). The second source is obtaining input from stakeholders, particularly people who are responsible for executing many of the tasks in a WBS. When displaying a WBS, a common format is a tree diagram rather than an outline. During initial construction, a WBS can be drafted using a white board or placing sticky notes on a large wall. Eventually, the WBS can be entered into a graphics package on a computer. An alternative display is to put it in typical outline format using a word processing or spreadsheet.

GOALS

- Encourage stakeholders to think hard about how to complete a project.
- Establish a solid groundwork to make realistic time and cost estimates.
- Build accountability among stakeholders.
- Force significant issues to arise early rather than later during a project.

OBSTACLES

- Being incomplete
- Not having consensus of stakeholders
- Not having specificity
- Not using the WBS in subsequent planning activities
- Not applying version control to the WBS

STEPS

1. Meet with project team members and client.
2. Identify the overall deliverable (e.g., product or service) to be built.

Work Breakdown Structure

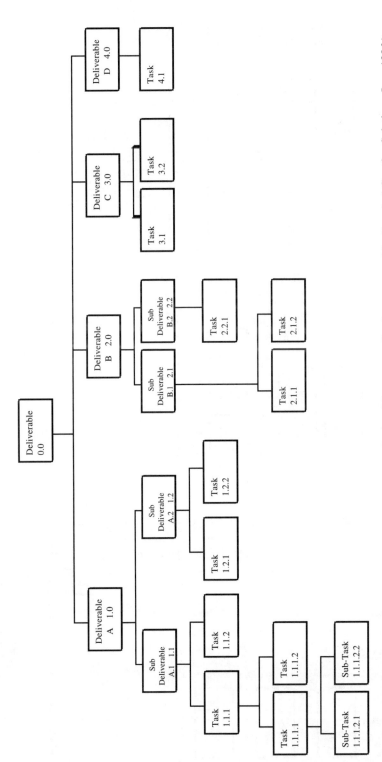

FIGURE 48 Work breakdown structure by deliverables. (From Project Management Seminar presented by Practical Creative Solutions, Inc., 1996.)

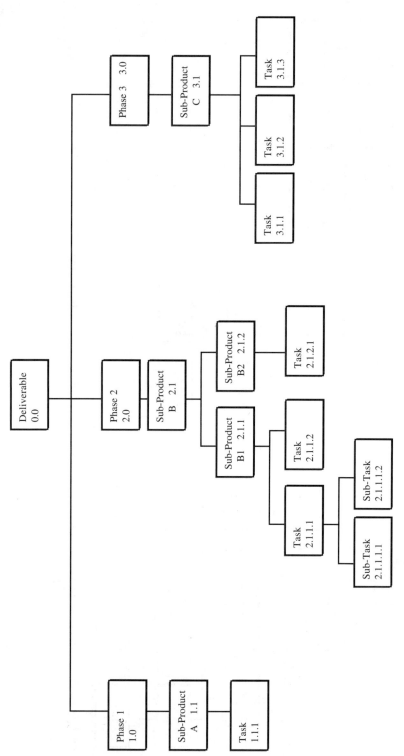

FIGURE 49 Work breakdown structure by phase. (From Project Management Seminar presented by Practical Creative Solutions, Inc., 1996.)

Work Breakdown Structure

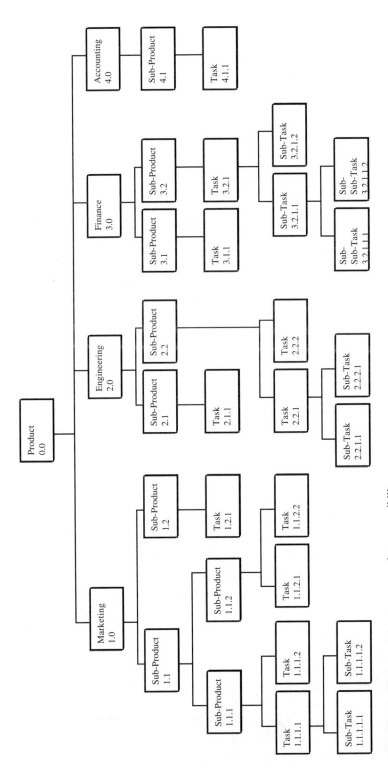

FIGURE 50 Work breakdown structure by responsibility.

3. Explode the overall deliverable down into varying level of details (e.g., subproducts).
4. Identify the tasks and subtasks for the lowest level of product details.
5. Verify that each item in the WBS has a unique numeric designator and a descriptive title.
6. Review the WBS with all relevant stakeholders to obtain consensus.
7. Place the WBS under version control upon reaching consensus.

WORKFLOW ANALYSIS

Workflow analysis is identifying and assessing the flow of information and application of controls in one or more processes. It requires looking at the as-is process (the way it currently exists) and developing ways to either improve it or replace it with a to-be process. Whether it is an as-is or to-be process, it consists of a set of activities, event constraints (e.g., rules, timing), and roles and responsibilities. Most processes contain a set of automated tools, or systems, that convert some type of input into some type of output. After an assessment is made, opportunities for improvement are identified, such as inaccurate information, lengthy cycle times, too many approvals, too much control, poor training, or poor quality. Workflow analysis is excellent for business process reengineering and for gaining an understanding of how a business process works.

GOALS

- Identify more effective and efficient ways of managing a project.
- Provide the basis for documenting procedures.
- Understand how the product or service will impact the customer's way of doing business.

OBSTACLES

- Inability to provide the time and other resources to perform workflow analysis
- Failure to perform an objective comparison between as-is and to-be processes
- Inability to ascertain the true cause of problems when conducting workflow analysis

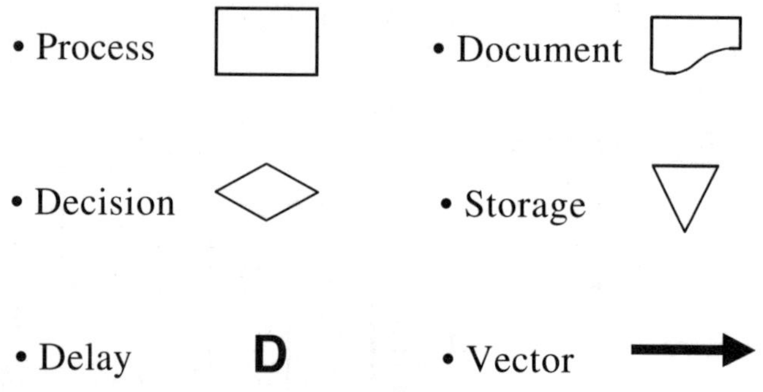

- Process   - Document

- Decision   - Storage

- Delay   D   - Vector

FIGURE 51 Workflow symbols.

STEPS

1. Identify the major activities of an as-is process.
2. For each as-is process, identify the fundamental rules, responsibilities, and constraints.
3. For each to-be alternative, identify the fundamental rules, responsibilities, and constraints.
4. Determine criteria to compare the as-is process with the to-be alternatives.
5. Select the preferred to-be process.
6. Develop a plan for implementing the new to-be process.

WORK FLOWS

A work flow is used to describe a process or procedure. It is often used to supplement or serve as an alternative to a narrative procedure. It represents flow of control; that is, it reflects the logical sequence of actions. Symbols are used to reflect these actions or items: process, decision, delay, document, storage, and vector (direction of flow).

GOALS

- Provide better clarity and conciseness over narrative procedures.
- Minimize the quantity of symbols to reduce complexity when describing a process or procedure.

OBSTACLES

- Inconsistent use of symbols
- Not providing a legend
- Developing the wrong type of flow diagram
- Too much clutter

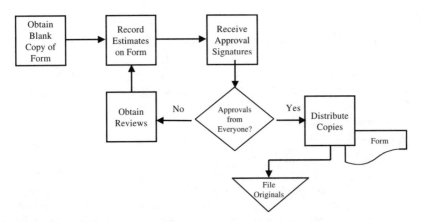

FIGURE 52 Example of work flow.

STEPS

1. Determine symbols.
2. Add legend.
3. Keep flowchart from becoming too cluttered.
4. Distribute (e.g., put in a project manual).
5. Update periodically.
6. Maintain under version control.

WORKPLACE DESIGN

Workplace design has a tremendous impact on the performance of a project team. The goal is to provide an environment that reduces impediments to intra- and interpersonal productivity. The physical aspects of workplace design deal with issues such as lighting, heating and ventilation, acoustics, ergonomics, and safety. The psychological aspects deal with issues such as privacy, boredom, fatigue, and spacing (from a cultural perspective). The workflow aspect deals with issues such as colocation, access to materials (including information), and layout.

GOALS

- Provide the opportunity to manage a project more efficiently and effectively.
- Increase morale and *esprit de corps*.

OBSTACLES

- Placing a low importance on workplace design
- Emphasizing aesthetics over workflow and location
- Failing to recognize the importance of the psychology behind workplace design

STEPS

1. Determine the flow of work on a project.
2. Arrange workstations to reflect the work flow.
3. Identify what conditions can positively or negatively impact productivity.
4. Remove the sources of negative productivity or alleviate their impact.
5. Recognize the psychological as well as physical impacts on productivity of poor workplace design.

WRITING

Writing is an effective means of communication when prepared clearly and concisely; however, most people on projects view writing as a necessary evil that gets very little attention unless requested of them. Often, writing is delegated to the worst performer on a team and is often overlooked when a project is under a time constraint.

The quality of most documentation reflects this circumstance. More often than not, documentation is replete with examples of poor grammar, misspellings, vague expressions, negativism, clichés, and jargon. Couple all of this with writer's block and it is amazing that anything gets written. The key to writing is to follow these actions: draft, edit, review, revise, approve, publish, and update. To ensure that people write well on a project, have an extra set of eyes review documentation for problems. In addition, ensure that documents, once published, are placed under version control and are accessible by everyone who must reference them.

Goals

- Communicate information clearly and concisely.
- Provide an audit trail regarding performance.
- Provide a form of knowledge management.
- Reduce the effects of turnover.

Obstacles

- Deemphasizing the importance of good writing
- Mixing active and passive voice inappropriately
- Assuming that the reader will get the idea despite poor writing

Steps

1. Identify the audience.
2. Determine the purpose.
3. Prepare the draft.
4. Either set the draft aside for a few days and then edit it or give it to someone else for an objective review.
5. Ensure that the writing addresses who, what, when, where, why, and how, if applicable.
6. After publication, store the document somewhere and place it under configuration control.

References

Ansoff, I.H., *Implanting Strategic Management*, Prentice Hall, Englewood Cliffs, NJ, 1984.
Armstrong, T., *Seven Kinds of Smarts*, Plume, New York, 1993.
Arnold, J.D., *The Complete Problem Solver*, John Wiley & Sons, New York, 1992.
Barr, J.T., *SPC Tools for Everyone*, ASQC Quality Press, Milwaukee, WI, 1993.
Bennis, W., *On Becoming a Leader*, Perseus Books, Reading, MA, 1989.
Berlack, H.R., *Software Configuration Management*, John Wiley & Sons, New York, 1992.
Blake, R.R. and McCanse, A.A., *Leadership Dilemmas: Grid Solutions*, Gulf Publishing, Houston, TX, 1991.
Block, P., *Stewardship*, Berrett-Koehler Publishers, San Francisco, CA, 1993.
Boar, B.H., *Application Prototyping*, John Wiley & Sons, New York, 1984.
Boar, B.H., *The Art of Strategic Planning for Information Technology*, John Wiley & Sons, New York, 1993.
Bolton, R., *People Skills*, Touchstone, New York, 1979.
Booch, G., *Object-Oriented Design with Applications*, Addison-Wesley, Reading, PA, 1994.
Bouldin, B.M., *Agents of Change*, Yourdon Press, Englewood Cliffs, NJ, 1989.
Bradford, D.L. and Cohen, A.R., *Managing for Excellence*, John Wiley & Sons, New York, 1984.
Bramson, R.M., *Coping with Difficult People*, Dell, New York, 1981.
Briner, W., Michael, G., and Hastings, C., *Project Leadership*, Gower, Brookfield, VT, 1990.
Brock, S.L., *Better Business Writing*, Crisp Publications, Menlo Park, CA, 1987.
Browning, T., *Capacity Planning for Computer Systems*, AP Professional, Boston, MA, 1995.
Brunetti, W.H., *Achieving Total Quality*, Quality Resources, White Plains, NY, 1993.
Buckley, F.J., *Implementing Configuration Management*, IEEE Press, New York, 1992.
Burley-Allen, M., *Listening: The Forgotten Skill*, John Wiley & Sons, New York, 1995.
Buzan, T., *Use Both Sides of Your Brain*, NAL/Dutton, New York, 1993.
Camp, R.C., *Benchmarking*, Quality Press, Milwaukee, WI, 1989.
Champy, J., *Reengineering Management*. Harper Business, New York, 1995.
Coad, P. and Yourdon, E., *Object-Oriented Analysis*, Prentice Hall, Englewood Cliffs, NJ, 1991.
Cornell, J.L. and Shafer, L.B., *Structured Rapid Prototyping*, Prentice Hall, Englewood Cliffs, NJ, 1989.
Correll, J.G. and Edison, N.W., *Gaining Control*, Oliver Wight Productions, Essex Junction, VT, 1990.
Covey, S.R., *The Seven Habits of Highly Effective People*, Fireside, New York, 1990.
Covey, S.R., *People-Centered Leadership*, Fireside, New York, 1992.
Crosby, P.B., *Quality Is Free*, McGraw-Hill, New York, 1979.
Crosby, P.B., *Completeness*, Dutton, New York, 1992.
Crosby, P.B., *Quality Is Still Free*, McGraw-Hill, New York, 1996.
Davis, L.N., *Planning, Conducting, and Evaluating Workshops*, University Associates, San Diego, CA, 1974.
De Bono, E., *Six Thinking Hats*, Little, Brown & Co., Boston, MA, 1985.
De Bono, E., *Practical Thinking*, Penguin Books, New York, 1991.
Demarco, T. and Lister, T. , *Peopleware*, Dorset House Publishing, New York, 1987.

Demarco, T., *Structured Analysis and System Specification*, Prentice Hall, Englewood Cliffs, NJ, 1979.
De Neuman, B., Ed., *Software Certification*, Elsevier Applied Science, London, 1989.
Doyle, D. and Straus, D., *How To Make Meetings Work*, Jove Books, New York, 1982.
Drake, J.D., *The Effective Interviewer*, AMACOM, New York, 1989.
Dunn, R.H., *Software Quality*, Prentice Hall, Englewood Cliffs, NJ, 1990.
Duyn, Van J., *The DP Professional's Guide to Writing Effective Technical Communications*, John Wiley & Sons, New York, 1982.
Estes, P.S., *Work Concepts for the Future*, Crisp Publications, Menlo Park, CA, 1996.
FitzGerald, J. and FitzGerald, A., *Fundamentals of Systems Analysis*, 3rd ed., John Wiley & Sons, New York, 1987.
Fleming, Q.W., *Cost/Schedule Control Systems Criteria*, Probus Publishing, Chicago, IL, 1992.
Fox, W.M., *Effective Group Problem Solving*, Jossey-Bass Publishers, San Francisco, CA, 1987.
Galitz, W.O., *Humanizing Office Automation*, QED Information Sciences, Wellesley, MA, 1984.
Gane, C. and Sarson, T., *Structured Systems Analysis: Tools and Techniques*, Prentice Hall, Englewood Cliffs, NJ, 1979.
Gardner, H., *Frames of Mind*, Basic Books, New York, 1993.
Gelinas, U.J., Oram, A.E., and Wiggins, W.P., *Accounting Information Systems*, South-Western Publishing, Cincinnati, OH, 1993.
Glass, R.L., *Software Communication Skills*, Prentice Hall, Englewood Cliffs, NJ, 1988.
Grady, R.B. and Caswell, D.L., *Software Metrics*, Prentice Hall, Englewood Cliffs, NJ, 1987.
Greenly, R.B., *How To Win Government Contracts*, Van Nostrand-Reinhold, New York, 1983.
Grose, V.L., *Managing Risk*, Prentice Hall, Englewood Cliffs, NJ, 1987.
Hammer, M. and Champy, J., *Reengineering the Corporation*, Harper Business, New York, 1994.
Hammer, M. and Stanton, S.A., *The Reengineering Revolution Handbook*, Harper Business, New York, 1994.
Harrington, H.J., *Business Process Improvement*, McGraw-Hill, New York, 1991.
Herrman, N., *The Creative Brain*, Brain Books, Luke Luce, NC, 1990.
Himmelfarb, P.A., *Survival of the Fittest*, Prentice Hall, Englewood Cliffs, NJ, 1992.
Holms, J.N., *Productive Speaking for Business and the Professions*, Allyn & Bacon, Boston, MA, 1969.
Humphrey, W.S., *Managing for Innovation*, Prentice Hall, Englewood Cliffs, NJ, 1987.
Hunt, V.D., *Reengineering*, Omneo, Essex Junction, VT, 1993.
Ishikawa, K., *What Is Total Quality Control?*, Prentice Hall, Englewood Cliffs, NJ, 1985.
James, M. and Jongeward, D., *Born To Win: Transactional Analysis with Gestalt Experiments*, Perseus Press, Portland, OR, 1996.
Janis, I.L., *Victims of Groupthink*, Houghton Mifflin, Boston, MA, 1972.
Jones, M.P., *What Every Programmer Should Know About Object-Oriented Design*, Dorset House, New York, 1996.
Juran, J.M., *Juran's Quality Control Handbook*, 4th ed., McGraw-Hill, New York, 1988.
Karrass, C.L., *The Negotiating Game*, Thomas Y. Crowell, New York, 1970.
Katzenbach, J.R., *Teams at the Top*, Harvard Business School Press, Boston, MA, 1998.
Katzenbach, J.R., *The Work of Teams*, Harvard Business Review, Boston, MA, 1998.
Katzenbach, J.R. and Smith, D.K., *The Wisdom of Teams*, Harvard Business School Press, Boston, MA, 1993.
Kerzner, H., *Project Management*, 5th ed., Van Nostrand-Reinhold, New York, 1995.

References

Kliem, R.L. and Ludin, I.S., *Just-in-Time Systems for Computing Environments*, Quorum Books, Westport, CT, 1994.
Kliem, R.L. and Ludin, I.S., *The People Side of Project Management*, Ashgate, Brookfield, VT, 1995.
Kliem, R.L. and Ludin, I.S., *Reducing Project Risk*, Ashgate, Brookfield, VT, 1997.
Kliem, R.L. and Ludin, I.S., *Project Management Practitioner's Handbook*, AMACOM, New York, 1998.
Kliem, R.L. and Ludin, I.S., *Managing Change in the Workplace*, HNB Publishing, New York, 1999.
Kliem, R.L. and Ludin, I.S., *Tools and Tips for Today's Project Manager*, Project Management Institute, Newtown Square, PA, 1999.
Kliem, R.L., Ludin, I.S., and Robertson, K.L., *Practical Project Management Methodology*, Marcel Dekker, New York, 1997.
Kobielus, J.G., *Workflow Strategies*, IDG Books, Foster City, CA, 1997.
Kouzes, J.M. and Posner, B.Z., *The Leadership Challenge*, Jossey-Bass, San Francisco, CA, 1988.
Kowal, J.A., *Behavior Models*, Prentice Hall, Englewood Cliffs, NJ, 1992.
MacKenzie, A., *The Time Trap*, AMACOM, New York, 1990.
McNain, C.J., and Leibfried, K.H.J., *Benchmarking*, Oliver Wight Publications, Essex Junction, VT, 1994.
Miller, W.C., *The Creative Edge*, Addison-Wesley, Reading, MA, 1986.
Murphy, M.A. and Xenia L.P., *Handbook of EDP Auditing*, 2nd ed., Warren, Gorham, and Lamont, Boston, MA, 1989.
Myers, G.J., *The Art of Software Testing*, John Wiley & Sons, New York, 1979.
Niebel, B.W., *Motion and Time Study*, Irwin, Homewood, IL, 1976.
Nierenberg, G.I., *The Art of Negotiating*, Cornerstone Library, New York, 1968.
Nierenberg, G.I., *The Art of Creative Thinking*, Cornerstone Library, New York, 1982.
Nirenberg, J.S., *Getting Through to People*, Englewood Cliffs, NJ, Prentice Hall, 1963.
Ohmae, K., *The Mind of the Strategist*, Penguin Books, New York, 1982.
Ohmae, K., *The Mind of the Strategist*, 2nd ed., Penguin Books, New York, 1983.
Palmer, H., *Enneagram*, Harper Collins, San Francisco, CA, 2001.
Preece, D.A., *Managing the Adoption of New Technology*, Routledge, London, 1989.
Rubin, T., *Overcoming Indecisiveness*, Avon Books, New York, 1986.
Rumbaugh, J. et al., *Object-Oriented Modeling and Design*, Prentice Hall, Englewood Cliffs, NJ, 1991.
Salton, G.J., *Organizational Engineering*, Professional Communications, Ann Arbor, MI, 1996.
Scherkenback, W.W., *The Deming Route to Quality and Productivity*, CEE Press Books, Washington, D.C., 1991.
Schultz, D.E., *Psychology and Industry Today*, 2nd ed., Macmillan Publishing, New York, 1978.
Taylor, D.A., *Object-Oriented Technology*, Servio, Alameda, CA, 1990.
VanAlstyne, J.S., *Professional and Technical Writing Strategies*, Prentice Hall, Englewood Cliffs, NJ, 1986.
VanGundy, A.B., *Creative Problem Solving*, Quorum Books, Wesport, CT, 1987.
Vollman, T.E., Berry, W.L., and Whybark, C.D., *Manufacturing Planning and Control Systems*, 3rd ed., Irwin, Homewood, IL, 1992.
Whitmore, J., *Coaching for Performance*, Pfeiffer & Company, Amsterdam, 1994.
Whitten, N., *Managing Software Development Projects*, John Wiley & Sons, New York, 1990.
Williams, L.V., *Teaching for the Two-Sided Mind*, Touchstone, New York, 1983.
Wonder, J. and Donovan, P., *Whole-Brain Thinking*, Ballantine Books, New York, 1984.